ABC
ORGANIC FARMING

ABC of Organic Farming

ABC ORGANIC FARMING

Amitava Rakshit
Department of Soil Science and Agricultural Chemistry,
Institute of Agricultural Sciences,
Banaras Hindu University,
Varanasi, 221005, India

H. B. Singh
Department of Mycology and Plant Pathology,
Institute of Agricultural Sciences,
Banaras Hindu University,
Varanasi, 221005, India

2015

JAIN BROTHERS (New Delhi)
16/873, East Park Road, Near Dr. N. C. Joshi Hospital,
Karol Bagh, New Delhi -110 005
Phones : 011- 23618426, 45097398
E-mail : jain_bros@hotmail.com
Website : www.thejainbrothers.com

Published by :

Shri Sunil Kumar Jain
for
Jain Brothers (New Delhi)
16/873, East Park Road, Karol Bagh, New Delhi-110 005
Phones : 011-23618426, 45097398

Branch Office at :

1. **M/s Sunil Book Service**
   Fatima Nagar, Opp. Andhra Bank,
   KAZIPET-506004 (A.P.)
   Ph. : 0870-2459617, Mob. : 9390115435
   E-mail : kpkulshrestha@yahoo.com
2. **M/s Jain Brothers**
   Vidya Vihar, Pilani (Rajasthan)
   Phone : 01596-242144

First edition : 2015

***Price* : ₹ 150.00**

ISBN : 978-81-8360-215-0

Laser Typesetting by :

Niyati *Graphics*
754/23, Khajoor Road,
Karol Bagh, New Delhi

Printed by :
BALAJI OFFSET PRINTERS, New Delhi-110035

प्रोफेसर पंजाब सिंह

Prof. Panjab Singh

FNAAS, FIAS, FNIE, FBRSI, FSEE
Ex Secretary, Dept. of Agriculture Research and Education (DARE),
and Director General, Indian Council of Agricultural Research (ICAR).
Ex VC (BHU, Varanasi)

Foreword

By the year 2050, Earth's population will double. If we continue with current farming practices, vast amounts of wilderness will be lost, millions of birds and billions of insects will die, and the public will lose billions of dollars as a consequence of environmental degradation. Clearly, there must be a better way to meet the need for increased food production. Organic farming is both dynamic and forward-looking but is also rooted in tradition. It is these traditions that can provide valuable starting points in debates over how organic farming should meet new challenges such as globalization, the emergence of new production techniques, and growing concern over equity and social justice in agriculture. There are a number of books and resources on how to produce organic crops. But in an easy to read style, ***ABC of Organic Farming*** is packed with incredibly useful information that is hard to find elsewhere. Significant and relevant issues pertaining to all aspects of organic farming have also been discussed in the book. The authors of the book have made an attempt to arrange different components of organic farming in a concise but comprehensive manner and deserve commendation for their painful efforts. This book is going to be an excellent resource material for farmers, and policy decision makers who want to support ecologically responsible farming practices for a sustainable future. I do hope that the present book will be a valuable source of information for students and teachers engaged in various applications of agriculture. I appreciate and congratulate the authors for bringing out highly useful and relevant publication of immense use.

(Panjab Singh)

Preface

Organic farming involves more than just farming without chemicals. It requires changes to many parts of the crop and livestock production system such as enhanced use of integrated pest, weed and nutrient management techniques. These include using such preventative measures as crop rotations, cover crops, improved genetics, optimum populations, stress management and sanitation. They are essential to improve plant and soil health since many chemical tools for pest and disease control are not used in organic production. Organic farmers choose not to use synthetic pesticides, fertilizers and antibiotics. It is very important to consider the transition to organic carefully. Making the transition too quickly can create financial hardship. During the first years of the transition yields will be decreased and there are few premiums for transitional organic products. Over time and with good management, profit levels should increase. Profits in organic agriculture will depend in part on the availability of market premiums. One should evaluate his reasons and goals for making changes. The transition to organic takes several years depending on the commodity and on the approach taken. The requirements for organic certification must be considered throughout the process of transition. The emerging sector of organic agriculture is described, and its potential is discussed in detail. Finally, information compiled show how traditional knowledge, social mobilization and agro-ecological approaches have been used to restore degraded natural resources while producing quality food. The small farmers who seek fully integrated food systems should be given recognition throughout the publication. They, along with the consumers who are creating market-based incentives for ecological management of agricultural systems, are at the centre of the attraction for this effort. The book further discusses the institutional support for nurturing existing knowledge and exchange in organic agriculture, support that could further enhance organic agriculture's positive impact on the natural and human environments without impairing the natural resource base. Our thanks to students, teachers and colleagues who have inspired us to plan for this book. We especially thank Prof. B.D. Singh, Prof. Bhagyaraj, Prof. K.B.R. Tilak, Prof. Ibrahim, Prof. Claassen, Prof. Subhendu Chaudhary who have made valuable suggestions and comments. Credits for illustrations, quotations from other individuals or publications are given with the individual items. No doubt further improvement of the book is also possible, and as such we shall greatly appreciate the suggestions and constructive criticism for its improvement.

Amitava Rakshit

HB Singh

Contents

CHAPTER 1

Introduction

1. The present reality

The growing human population above 1.8% per year and the intensive agriculture practices with the use of synthetic agro-chemicals like fertilizers, pesticides etc., is exerting immense pressure globally on the natural resources used for farming. In the last decade a series of questions are raised which warrants attention in order to produce food, fodder, fibre, fuel and other agricultural commodities to the burgeoning population without deteriorating the natural resources. Some of the concerns are:

- The cost of intensive agriculture (15,000-18000 mega joules/ha as compared to 8,000-10,000 mega joules /ha in organic agriculture) has eroded the biological potential of the soil and depleted the underground water resources.
- In intensively cropped areas, more energy (≈ 78 MJ/kg of NPK) in the form of mineral fertilizers, chemical pesticides and farm machinery is required each year to produce the same quantity of grain, or other farm products.
- Increase in damage caused by pests and disease (≈ 15-25%) worsening.
- Scientific studies reveal that the modern agricultural system increases the volume of the 'green house gases' (≈1.37 tonnes of green house gases emissions per capita) in the atmosphere.
- Genetic heterogeneity of earlier agricultural systems is replaced by genetic homogeneity in crops and animals, thereby making them more vulnerable to biotic and abiotic stresses.
- Large subsidies provided in developed and developing countries (≈ 100 crores in input sector; input subsidy is equal to 10% of the values of the total agricultural output for India) to encourage farmers to increase productivity.
- Residual toxicity in the food chain and increasing concerns on human health and nutrition.

- Sustainability of agriculture in consonance with economics, ecology, equity, energy and socio-cultural dimension.

These apprehensions have led to a concept in agriculture where technology development has to pay due attention to ecological sustainability along with economic efficiency. Sustainable or alternative agriculture is a revised type of agriculture, rich in technology and information and less intensive in the use of energy and market purchased inputs rather concentrating on the local available resources. Sustainable agriculture is a unifying concept, which considers ecological, environmental, philosophical, ethical and social impacts, balanced with cost effectiveness and practicality. In other words, an earth friendly system of farming which is ecologically sound, economically viable socially responsible and politically feasible, is the need of the hour.

1.1 Sustainable Agriculture

Sustainability is defined considering different dimensions, ranging from the narrow focus on economics or production to the incorporation of culture and ecology in broader perspective. In scientific illustrations sustainability is usually understood in a more holistic way. Basically ssustainable development is an organising principle for human life on a finite planet. It posits a desirable future state for human societies in which living conditions and resource-use meet human needs without undermining the sustainability of natural systems and the environment, so that future generations may also have their needs met. Several national and international organizations (like ICAR, ICRISAT, IUCN, UNEP, WWF) defines sustainable development as improving the quality of human life while living within the carrying capacity of supporting ecosystems. This necessitates the conservation and improvement of natural ecosystems through research, training, technology, community co-operation and public policies. The World Bank describes the requirement for sustainable agricultural systems as being "environmentally sound, financially and economically feasible and socially acceptable". During the earth summit in Rio de Janeiro in 1992, the UN Food and Agriculture Organization defined sustainable and rural development as follows: 'Sustainable development is the management and conservation of the natural resource base and the orientation of technological and institutional change in such a manner as to ensure the attainment and continued satisfaction of human needs for present and future generations. Such sustainable development in agriculture, forestry and fisheries sectors conserves land, water, plant and animal genetic resources, is environmentally non-degrading, technically appropriate, economically viable and socially acceptable'.

Sustainable agriculture should not be understood as a specific agricultural practice, technology or system rather a societal comprehensive goal to be pursued forever and for everyone and guided by general principles like:

1. Principle of practical use of renewable resources

A system which depends on exhaustible resources such as fossil fuels cannot be sustained indefinitely. A sustainable system would use renewable energy sources such as biological, geothermal, hydroelectric, solar, or wind. Use of recyclable resources such as groundwater at rates greater than recharge depletes reserves and cannot be sustained.

2. Principle of protecting the consistency of natural systems

Our current thinking focuses on reducing the rate of degradation of natural and agricultural ecosystems. A system will not be sustainable as long as the goal is simply to decrease the rate of its deterioration. Sustainable agricultural systems should maintain or improve groundwater and surface water (conjunctive use of water) quality and regenerate healthy agricultural soils.

3. Principle of improving the quality of life of individuals and group of individuals

In order to slow down rural to urban migration, rural communities must offer people a good standard of living including diverse employment opportunities, health care, education, social services and cultural activities. Young people must be afforded opportunities to develop rural enterprises, including farming, in ways which care for the land so that it may be passed onto forth coming future generations in as good or in better condition than it was received.

4. Principle of financial benefit

Transition to new ways of knowing, doing and being require incentives for all participants. Some of these incentives are necessarily economic gains. Systems and practices that do not include profitability as one of the prime motivators will not be voluntarily implemented.

5. Principle of morally correct

Holistic or whole-system analysis views an agroecosystem as a dynamic community of soil, water, air and biotic species. All parts are important because they contribute to the whole. This ethic strives to protect the health of the land community, that is its capacity for self-renewal.

In 1995, FAO defined sustainable agriculture and rural development more specifically as a process that meets the following criteria:

- Ensures that the basic nutritional requirements of present and future genera tions, qualitatively and quantitatively.
- Provides durable employment, sufficient income and decent living and working.
- Maintains and enhances the productive capacity of the natural re source base as a whole and the regenerative capacity of renewable resources, without disrupting the functioning of basic ecological cycles and natural balances, destroying the socio-cultural attributes of rural communities, or causing contamination of the environment.
- Reduces the vulnerability of the agricultural sector to adverse natural and socioeconomic factors and other risks, and strengthens self-reliance'.

Although sustainable agriculture can be defined in many ways, but ultimately it seeks to sustain agrarian, resources and communities by promoting farming practices and methods that are profitable, environmentally sound and good for communities. Sustainable agriculture fits into and complements modern agriculture. It rewards the true values of producers and their products.

CHAPTER 2

Sustainable Agriculture

2. Sustainable Agriculture is the Answer

With the population exceeding 6.7 billion and growing by over 6 million a month, the need to protect agricultural land and to increase food production has become critical. Among the viable alternatives, sustainable agriculture have the answer. The importance of sustainable agriculture in comparison to the conventional agriculture can be envisioned through three dimensions, ecological, economic and social (Figure 1).

2.1 Ecological Sustainability

Conventional agriculture is not ecologically sustainable as the usage of lot of fossil energy has a direct impact on the natural resources, reducing soil fertility, causing soil erosion and contributing to the global climatic change. The principles and practices of sustainable agriculture are ecologically sound and has many edges with reference to the following points:

Soil fertility

The use of synthetic fertilizers and pesticides has a detrimental impact on the fertility of the soils. Sustainable agriculture improves the fertility of the soils by promoting the development of earthworms and soil flora and fauna. A gram of healthy soil can contain some 600 million microorganisms and tens of thousands of different species of bacteria and fungi as well as organic and inorganic matter, air, water and water vapour. The utilization of the crop residues and weeds into soil through appropriate composting technologies improves the organic matter and avoids acidification, improve soil structure, percolation and soil aggregate stability which in turn prevents soil deterioration.

Water

The main threats posed by conventional agriculture on quality of water are, use of fertilizers, lack of protective soil cover, narrow crop rotations and frequent tillage, high levels of available nitrogen after harvest and contamination of water by application of synthetic pesticides. Sustainable agriculture practices improve the organic matter of the soil which in turn enhances the percolation and retention of the ground water resulting in increased ground water recharge thereby reducing run-offs.

Biodiversity

The major reason for the destruction of the biodiversity is the conversion of multifunctional, integrated ecosystems into a economy driven monoculture system is practiced. It not only reduces the crop diversity but the numerous varieties of an individual crop reduce drastically. In addition, the use of synthetic agrochemicals wipes off the indigenous population of organism such as mycorrhizal fungi, actinomycetes and thus disturbs the nutrient cycling of the soil eco-system. Sustainable agriculture rejuvenates the biodiversity in agro-ecosystems.

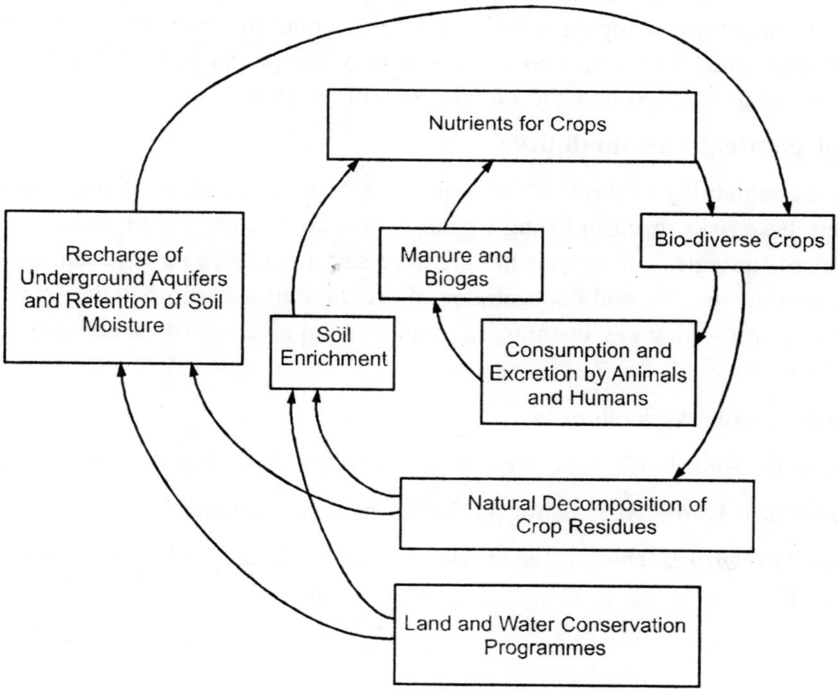

Figure 1: Enigma of Sustainable agriculture

Pollution

Conventional agriculture pollutes the ecology and the food, leading to a series of health hazards and pollution problems. Such hazardous chemicals are not used in sustainable agriculture, instead various plant based products, biological and agronomic practices are adopted to manage the pests.

Landscape

The rural landscape is clad with agriculture and forestry. Non scientific use of landscape causes erosion, landslides and flooding, clogs irrigation channels and reduces the ability of land to support the local habitat. Impoverished rural people migrate to the cities creating slums which further destroy the landscape. Rehabilitating ecologically damaged areas needs huge investments from both public and private sector which few countries can afford. Sustainable agriculture avoids these problems by improving productivity, conserving the soil, avoiding the expansion of farming into unsuitable areas and preserving and creating rural jobs.

Climate

The modern hitech agriculture contributes significantly to the global climatic change. Such impacts can be reduced significantly by adopting sustainable agriculture.

2.2 Economic Sustainability

The long term economic viability is vital for the sustainability of agriculture. In addition to strategies for preserving natural resources and changing production practices, sustainable agriculture requires a commitment to changing public policies, economic institutions, and social values. Strategies for change must take into account the complex, reciprocal and ever-changing relationship between agricultural production and the broader flexible society.

2.3 Social and political Sustainability

The social sustainability of farming techniques is related to the ideas of social acceptability and justice. Ignoring these risks result in losing valuable local knowledge and provoking political unrest. A wide diversity of strategies and approaches are necessary to create a more sustainable food system. These will range from specific and concentrated efforts to alter specific policies or practices, to the longer-term tasks of reforming key institutions, refining economic priorities, and challenging widely-held social values.

In short Sustainable Agriculture is:

- ***Ecologically Sound:*** We must preserve the resource base that sustains us all.
- ***Economically Viable:*** If it is not profitable, it is not sustainable.
- ***Socially Supportive:*** The quality of life of farmers, farm families and farm communities is important.

CHAPTER 3

Organic Agriculture

3. Background of Organic Agriculture

Organic agriculture is the oldest form of agriculture on earth. Farming without the use of fossil fuel-based chemicals was the sole option for farmers until post-World War II (WWII). The war brought with it technologies that were useful for agricultural production. For example, ammonium nitrate used for munitions during WW II evolved into ammonium nitrate fertilizer; organophosphate nerve gas production led to the development of powerful insecticides. These technical advances since WWII have resulted in significant economic benefits as well as environmental and social detriments. Organic agriculture seeks to utilize those advances that consistently yield benefits while discarding those methods that have led to negative impacts on society and the environment, such as pesticide pollution and insect pest resistance. Instead of using synthetic fertilizers and pesticides, organic farmers utilize crop rotations, cover crops, and natural-based products to maintain or enhance soil fertility. In the present day perspective among the various alternatives, organic farming is gaining acceptance throughout the globe as it has a potential to provide pragmatic solution to mitigate the sequel of maladies of conventional modern farming. In Asia, the Hindu, Buddhist and Jain monks visualized the sustainability of agriculture through the non-violent approach, before facing the onslaught of chemical farming, which is apt at present. Organic agriculture is getting popular globally due to the unprecedented demand for natural foods in the developed world and growing awareness regarding the health and environmental hazards related to the conventional chemical agriculture. Decline in food production, degeneration in native soil fertility and deterioration in environmental quality have emerged as the biggest problems facing global agriculture today. The tremendous demand for organically produced food at the international market is evident from the growing retail sales of the organic products in the developed countries. Presently 37 million hectares of agricultural land are managed organically by almost 1.4 million producers. The countries with the highest numbers of producers are India (>350000 producers), Uganda (180000) and Mexico (130000). More than one third of organic producers are in Africa.

The world organic products market has seen sustained growth over the past decade and has reached an estimated total value of US$50 billion.U.S, Europe and Japan are the biggest markets for the organically grown products. The retail sales of organic products in US ,Europe and Asia are to the tune of $ 8 billion and $ 8.4 billion respectively. Japan is also a major importer of organic products. The growing awareness on healthy food and increasing demand of organic products in the international market has led to the creation of new export avenues for the developing countries. Organic farming is a production system, which excludes the use of synthetically compounded inputs. It relies on organic manures produced from farm wastes and other biomass. It also encompasses assembly of various techniques and practices like intercropping, mulching, cover cropping, trap cropping etc. Organic farming also employs various biological pest control methods, which eliminates the use of synthetic chemicals even at the storage levels. A thorough understanding of agro ecological parameters of the locality is required to make organic farming a sustainable and feasible production system. Eventually, this helps to adopt the locally suitable methodologies with a proper and appropriate combination of various resources available on the farm. Being a holistic production management system, organic farming will promote and enhance environmental quality including biogeochemical cycles and soil biodiversity. The hassle is on improving the on-farm management rather than off-farm external inputs. Though there is a growing demand for safe food products in the international market, the aim of organic agriculture is not exports alone. It has a great role in improving the livelihoods by reducing the cost of production of the poor small and marginal farmers who have small land holdings. Fertilizers and pesticides have become major cost of production along with the cost of other input like seeds, and as a result of which farming is not economically feasible. The only option left is to identify low cost agriculture technologies which not only reduce the cost of production but also brings in sustainability to agriculture. The most important factor in low cost agriculture is that all the inputs have to be prepared on the farm and the off-farm inputs have to be reduced to the extent possible.

3.1 Reduced cost Organic Agriculture Systems

Different low-cost organic agriculture systems are being practiced all over the world. Some of the systems which are most popular are mentioned below.

3.1.1 Vedic Agriculture

Vedic agriculture pertains to the knowledge of farming which is presented in the age old Indian scripture the 'Vedas' about 10,000 years ago. The ancient Indian culture taught to venerate the earth as mother, sky as father, air as prana (soul), the sun as energy and the water streams as life sustaining veins. This system of farming makes use of various forms of inputs and it is not just some preparations made for providing nutrition or managing problems. There is a lot of significance of the Panchamahabhutas *i.e.* the five basic elements *prithvi* (Earth), *aap* (Water), *tejas* (Fire or light), *vayu* (Air) and *akasha* (Space). Bhutas means the living creatures. Mahabhutas are prior and superior to living beings since the very existence of the latter is totally dependent on them. The Mahabhutas are the primary natural resources essential for all agriculture. These Panchamahabhutas have a significant effect on plants. In Vedic agriculture, all the activities of farming are prescribed to be done at a specific time. The time bound nature of activities helps to harness the maximum positive effect of the Panchamahabhutas. This can be achieved by synchronizing all the activities as per the Panchang or the planting calendar. Planting calendar is a very important and a no cost input which can be easily adopted. Second emphasis is on the importance of Cow in vedic agriculture. It is said in the Indian scriptures that all the deities reside in different body parts of the cow. A variety of preparations can be made from the cow dung, cow's urine, cow's ghee, cow's milk, and curd or yoghurt made out of cow's milk. In addition, a variety of preparations for nutrition, pest and disease management are made by utilizing various medicinal herbs and other plants. Medicinal herbs like Shatavari (*Asparagus spp.*), Bhallataka (*Semecarpus anacardium*), Ashwagandha (Withania somnifera), Kana (Piper Longum), Guggulu (*Comniphora mukul*), Asana (*Terminalia tomentosa*), Shatapushpa (Anethum sowa), Triphala (Amalaki (*Emblica officinalis*), Bibhitaki (*Terminalia bellirica*), and Haritaki (*Terminalia chebula*)), Karanja (*Pongamia glabra*), Adathoda (*Adhatoda vasica*) etc. and other plants like Haridra (*Turmeric*), Ginger (*Zingiber officinale*), Oleander, Karavira (*Nerium*), Tulasi (*Ocimmum spp.*), Tobacco (*Nicotiana tabacum*), Langli (*Gloriosa superba*), Arka (*Calotropis*) and Sami (*Prosopis spp.*) are used in plant protection.

3.1.2 Organic Agriculture

Organic agriculture as widely known also involves the basic cultural practices of farming and the use of certain inputs for providing mineral nutrition and protection to the plants. ICAR defines organic farming as a system that is designed to produce agricultural products that relies on techniques such as crop rotation, green manure, compost, and biological pest control and substances that maintain the integrity of organic agricultural products until they reach the consumer. These practices help to manage the problems of pests to a larger extent. In addition, the inputs used in organic farming are:

Vermicompost

Vermicompost serves a dual benefit of providing good quality manure with natural nutrients for the soil and plants as well as solve the problem of wastes at the farm.

Biofertilizers

A biofertilizer is a substance which contains living microorganisms which, when applied to seed, plant surfaces, or soil, colonizes the rhizosphere or the interior of the plant and promotes growth by providing nutrients to the soil and plants by converting the nutrients from the non-available to the available form by the biological process of nitrogen fixation and phosphate solubilization The microorganisms widely used singly and in combinations for these purposes are *Azotobacter, Azospirilium* and Blue green algae, *Azolla* etc. These can meet the nitrogen requirements of the crops to a larger extent. However, there are many other microbes which mobilize the nutrients and improve the quality of soils by different ways and their effi ciencies should be tapped under sustainable agriculture.

Bio pesticides

Biopesticides, a class of biological pesticides, include several types of pest management intervention: through predatory, parasitic, or chemical relationships. Biopesticides consists of microbial pesticides and biochemical pesticides or herbal pesticide. Botanicals or herbal are the preparations made from specific plants for managing the pest and disease problems of the plants. A wide base of a variety of plants exist, different parts of which can be utilized in making the botanicals. Among them the most prominent are *Adathoda zeylanica, Adathoda vasica, Agave americana, Allium sativum, Anacardium occidentale, Argemone mexicana, Azadirachta indica, Brassica campestris, Capsicum frutescens, Cassia tora, Cinnamomum verum, Ocimum sanctum, Parthenium bysterophorus, Sesamum indicum, Annona squamosa, Melia azadiract, Hardwikia binata* etc.

3.1.3 Biodynamic Agriculture

Biodynamic Agriculture is a holistic system of farming developed by Dr. Rudolf Steiner, a scientist, philosopher and founder of the Waldorf School in Europe in 1924. The concept is as the word itself would suggest: a combination of biological and dynamic practices, "bio" meaning life and "dyn" meaning force. Like organic farming, biodynamic farms stress biological methods in regard to humane treatment of animals, food quality and soil health .This system of farming is based on the lines of traditional systems being practiced in India since ages. It is based on systematic and synergistic harnessing of energies from cosmos, earth, plant and cow. It includes certain dynamic practices in addition to the basic biological practices of cultivation. The inputs utilized in this system are:

i) Biodynamic (BD) preparations numbered from 500-508

ii) Liquid manures

iii) Cow Pat Pit manure (CPP)

iv) Biodynamic stem paste

The biodynamic preparations are numbered from 500 to 508. Preparation 500 and 501 are made by using cow dung and quartz silica. The other seven preparations are made by fermenting parts of certain herbs in a peculiar manner. These herbs are yarrow, chamomile, stinging nettle, oak bark, dandelion, valerian and equisetium respectively. These preparations are called the compost preparations and are utilized in making the biodynamic compost. These preparations activate synthesis of different nutrients in the final biodynamic compost. These are the components of biological agriculture, capable of affording long-term sustainability to agriculture and particularly to the ecosystem. Other preparations like liquid manures, CPP and biodynamic stem paste are utilized for a number of cultural practices in farming. Another unique feature in biodynamic agriculture system is the use of biodynamic planting calendar for various farm operations. Efforts are made to enliven the soil and restore its fertility in the form of humus, increase the living system of soil by skilful application of appropriate crop rotation.

3.1.4 Permaculture

Permaculture is a branch of ecological design, ecological engineering, and environmental design that develops sustainable architecture and self-maintained agricultural systems modeled from natural ecosystems. The term permaculture was first coined by Bill Mollison an Australian ecologist, and one of his students, David Holmgren. It is a land use and community building movement which strives for the harmonious integration of human dwellings, microclimate, annual and perennial plants, animals, soils, and water into stable, productive communities(Figure 2). The focus is not on these elements themselves, but rather on the relationships created among them by the way we place them in the landscape. This synergy is further enhanced by mimicking patterns found in nature . Permaculture design emphasizes patterns of landscape, function, and species assemblies. It determines where these elements should be placed so they can provide maximum benefit to the local environment. The central concept of permaculture is maximizing useful connections between components and synergy of the final design. The focus of permaculture, therefore, is not on each separate element, but rather on the relationships created among elements by the way they are placed together; the whole becoming greater than the sum of its parts. Permaculture design therefore seeks to minimize waste, human labor, and energy input by building systems with maximal benefits between design elements to achieve a high level of synergy. As such, permaculture design concepts are applicable to urban as well as rural settings, and are appropriate for single households as well as whole farms and villages.

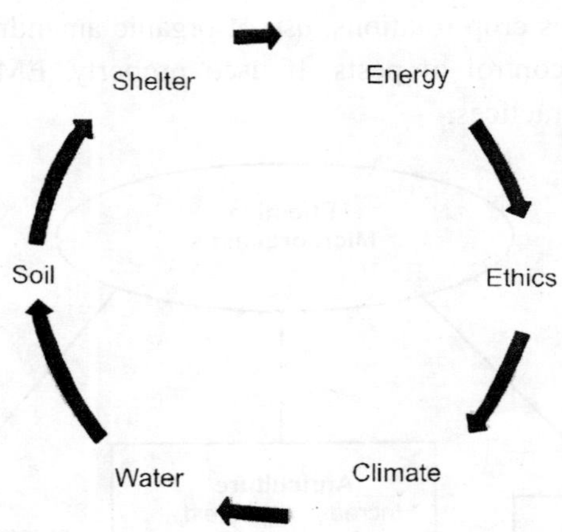

Figure 2: Different components of Permaculture

3.1.5 Effective Microorganisms (EM)

For many years, soil microbiologists and microbial ecologists have described soil microorganisms as beneficial or harmful according to their functions and how they affect soil quality, plant growth and yield, and plant health. Beneficial microorganisms are those that can fix atmospheric nitrogen, decompose organic wastes and residues, detoxify pesticides, suppress plant diseases and soil-borne pathogens, enhance nutrient cycling, and produce bioactive compounds such as vitamins, hormones and enzymes that stimulate plant growth. Harmful microorganisms are those that can induce plant diseases, stimulate soil-borne pathogens, immobilize nutrients, and produce toxic and putrescent substances that adversely affect plant growth and health. A more specific classification of beneficial microorganisms has been suggested by Higa which he refer to as "Effective Microorganisms" or EM.

The concept of EM was developed by Professor Teruo Higa, University of the Ryukyus, Okinawa, Japan. He reported in the 1980s that a combination of approximately 80 different microorganisms is capable of positively influencing decomposing organic matter such that it reverts into a "life promoting" process. Higa invokes a "dominance principle" to explain the effects of his EM. He claims that three groups of microorganisms exist: microorganisms involved in regeneration, microorganisms involved in decomposition, degeneration and opportunist microorganisms. EM consists of mixed cultures of beneficial and naturally-occurring microorganisms that can be applied as inoculants to increase the microbial diversity of soils and plant. Research has shown that the inoculation of EM cultures to the soil/plant ecosystem can improve soil quality, soil health, and growth, yield, and quality of crops (Figure 3). EM contains selected species of microorganisms including predominant populations of lactic acid bacteria and yeasts and smaller numbers of photosynthetic bacteria, actinomycetes and other types of organisms. All of these are mutually compatible with one another and can coexist in liquid culture. EM is not a substitute for other management practices. It is, however, an added dimension for optimizing our best soil and crop management practices such as crop rotations, use of organic amendments, conservation tillage, crop residue recycling, and bio control of pests. If used properly, EM can significantly enhance the beneficial effects of these practices.

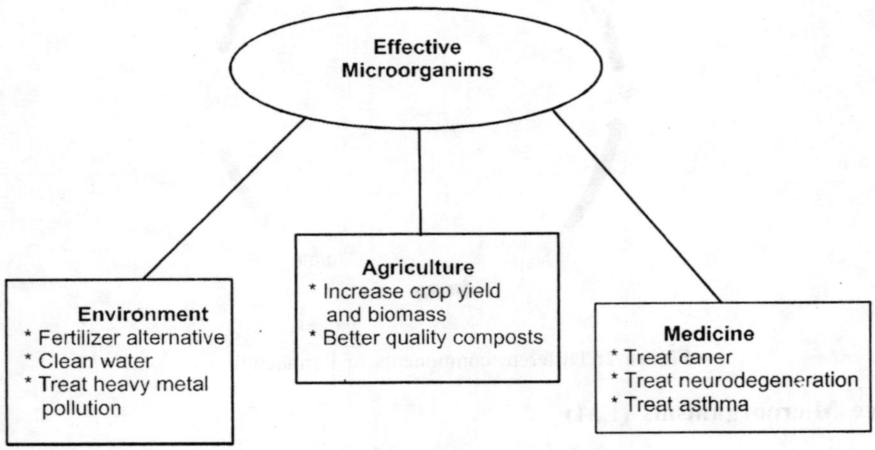

Figure 3: Different components of Effective microorganisms

3.1.5 Natural Farming

The philosophy and practice of Natural Farming (Figure 4) arose from Mokichi Okada's research, in the late 1930's and 1940's. Masanobu Fukuoka (1913–2008) was a Japanese farmer, ecologist and philosopher who described his way of farming as *shizen nôh* in Japanese. He discovered that if our agricultural methods respect the soil and enhance its innate power, then our food will always be rich in natural energy and will nourish and support our health. Human beings can thrive only by living in harmony with Nature. Okada discovered that everything on Earth, including vegetables and fruits, has a visible, physical side and an invisible, spiritual side. Healthy uncontaminated crops have an abundance of this invisible energy, which gives more energy to those who consume the crops.Unlike some of the practices of organic farming, Nature Farming holds this view of Nature as fundamental.Natural farming can also be described as ecological farming and is related to fertility farming, organic farming, sustainable agriculture, agroforestry, ecoagriculture and permaculture but should be distinguished from biodynamic agriculture.

The four basic tenets to natural farming are:

a) ***No cultivation*** - that is, no ploughing or turning the soil, the earth should be left to cultivate itself naturally by the penetration of plant roots, and the activity of micro-organisms and earthworms.

b) ***No chemical fertilisers*** - left alone the soil can maintain its own fertility through the cycle of plant and animal life. The only manure used is a light layer of poultry manure on top of the straw and clover mulch and this is usually provided by ducks and chickens roaming free.

c) ***No weeding by tillage or herbicides*** – weeds play a part in building soil fertility and should be controlled by mulch and other natural means rather than eliminated.

d) ***No pesticides*** – the best approach to disease and insect control is to grow sturdy crops in a healthy environment. Natural predators should be encouraged to take care of any unwanted pests.

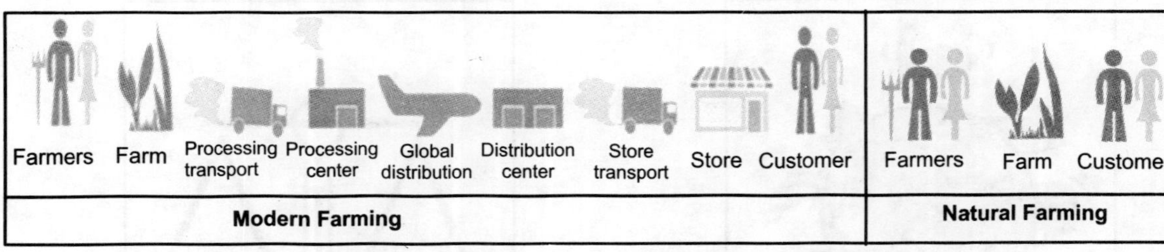

Figure 4: Different components of Modern farming and Natural Farming.

3.1 Principles of Organic Agriculture

The general principles of organic agriculture include the following:

1) Principle of Health

The health of the human beings and communities cannot be separated from the health of the ecosystems. Always healthy soils produce healthy crops and create healthy environment.

2) Principle of Ecology

Organic agriculture should attain ecological balance through the design of farming systems, establishment of habitats and maintenance of genetic and agricultural diversity. The management practices must be adapted to the local conditions, ecology, culture and scale.

3) Principle of Fairness

Fairness is characterized by equity, respect, justice and stewardship of the shared world both among people and in their relations to other living beings. Organic agriculture should provide food sovereignty, reduction of poverty and good quality of life to all.

4) Principle of Care

Precaution and responsibility are the key concerns in management, development and technology choices in organic agriculture. Along with scientific understanding practical experience, traditional wisdom, indigenous knowledge which are time tested provide valid solutions to ensure agriculture is healthy, safe and ecologically sound.

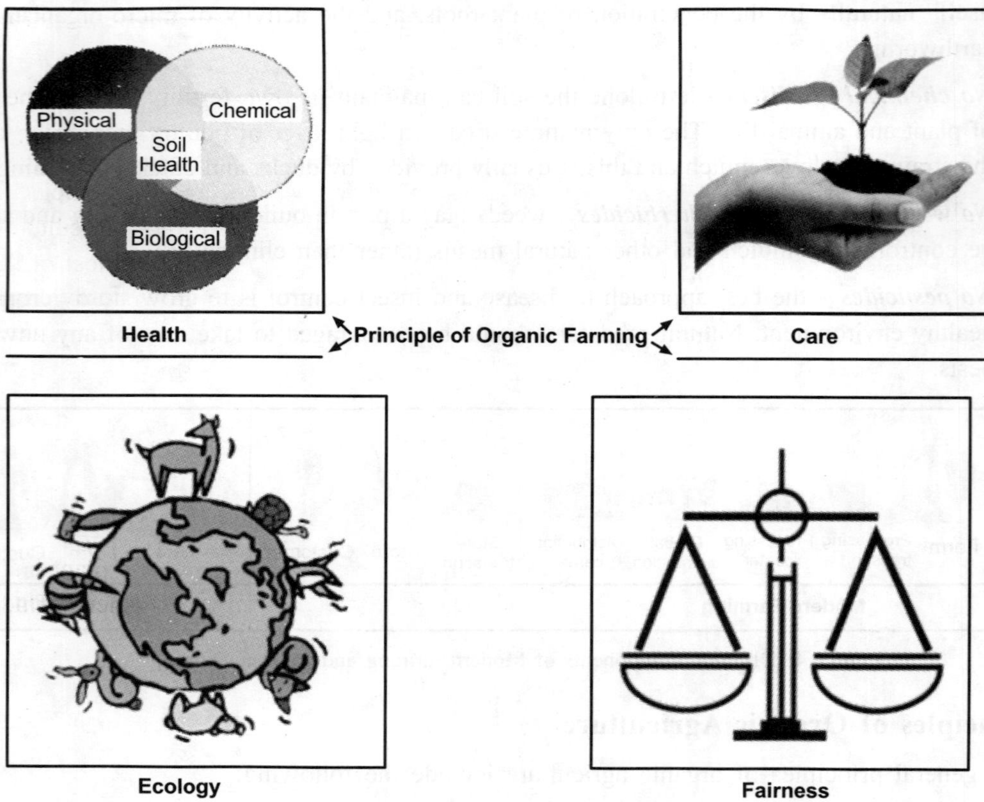

Figure 5: Broad principles of organic farming

All the four broad principles (Figure 5) are befitting the rich culture and heritage of India . The above principles of organic farming are achieved by :

1. Protect the environment, minimize soil degradation and erosion, decrease pollution, optimize biological productivity and support a sound state of health.
2. Maintain long-term soil fertility by optimizing conditions for biological activity within the soil.
3. Maintain biological diversity within the system.
4. Recycle materials and resources to the greatest extent possible within the enterprise.
5. Provide attentive care that promotes the health and meets the behavioral needs of livestock.

6. Prepare organic products, emphasizing careful processing, and handling methods in order to maintain the organic integrity and vital qualities of the products at all stages of production.
7. Rely on renewable resources in locally organized agricultural systems.

3.3 Why to advocate organic farming for small farmers?

Some of the advantages of organic farming for small farmers:

- **High premium:** Organic food is normally priced 20 - 30% higher than conventional food. This premium is very important for a small farmer whose income is just sufficient to feed his/her family with one meal.
- **Low investment:** Organic farming normally does not involve capital investment as high as that required in chemical farming. Further, since organic fertilizers and pesticides can be produced locally, the yearly costs incurred by the farmer are also low. Agriculture greatly depends on external factors such as climate, pests, disease. Furthermore, most of the small farmers are dependent on natural rain for water. Therefore in cases of natural calamity, pest or disease attack, and irregular rainfall, when there is a crop failure, small farmers practicing organic farming have to suffer less as their investments are low.
- **Less dependence on money lenders:** Many small farmers worldwide commit suicide due to increasing debt. Since chemical inputs, which are very costly, are not required in organic farming, small farmers are not dependent on money lenders. Crop failure, therefore, does not leave an organic farmer into enormous debt, and does not force him to take an extreme step.
- **Synergy with life forms:** Organic farming involves synergy with various plant and animal life forms. Small farmers are able to understand this synergy easily and hence find it easy to implement them.
- **Traditional knowledge:** Small farmers have abundance of traditional knowledge with them and within their community.Most of this traditional knowledge cannot be used for chemical farming. However, when it comes to organic farming, the farmers can make use of the traditional knowledge. Further, in case of organic farming, small farmers are not dependent on those who provide chemical know-how.

CHAPTER 4

Standards, Regulations and Certifications

4. Standards, Regulations and Certification

To protect the interest of consumers regarding the quality of organic farm produce some regulatory measures have been adopted in production and marketing of such products. These measures ensure quality production of the products. Certain regulatory measures are developed by the governments while others by the consumers or farmer federations. Some of the regulatory mechanisms which have become very popular are furnished below.

4.1 Good Agricultural Practices (GAP)

The increasing awareness of the deleterious effects of indiscriminate use of artificial inputs in agriculture such as chemical fertilizers and pesticides has led to the adoption of organic farming as an alternative method for conventional farming. Farmers are adopting organic farming due to the advantages like:

- It is self sustaining and socially and ecologically superior over the conventional farming.
- It makes use of cost effective management practices involving the use of farm inputs produced within the farm.
- It is less expensive.
- It is environment friendly as there is no pollution of soil, water and air.
- It enriches the soil and the local ecology.
- The produce is free of contamination from chemical residues has better taste, flavour and nutritional value.
- Seeds have more vitality and are good for successive generations.
- Growing markets for safe food.

It is prerequisite to understand and follow GAP before adopting organic agriculture. This would help the farmer to plan and prepare the needed inputs by efficiently utilizing the local resources. The concept of GAP has evolved in recent years in the context of a rapidly changing and globalizing food economy. In addition, the concerns and commitments of a wide range of stake holders on food production and security, food safety and quality and the environmental sustainability of agriculture have also promoted the concept of GAP. These stakeholders include governments, food processing and retailing industries, farmers and consumers.

GAPs are a set of principles, regulations and technical recommendations applicable to production, processing and food transport, addressing human health care, environment protection and improvement of worker conditions and their families (Figure 6). According to the Food and Agriculture Organization (FAO), GAP is the application of available knowledge to address environmental, economic and social sustainability for on-production and post-production process resulting in production of safe and healthy food and non-food agricultural products. Many farmers in developed and developing countries already apply GAP through sustainable agricultural methods such as integrated pest management, integrated nutrient management and conservation agriculture. These methods are applied in a range of farming systems and scales of production units, facilitated by supportive government policies.

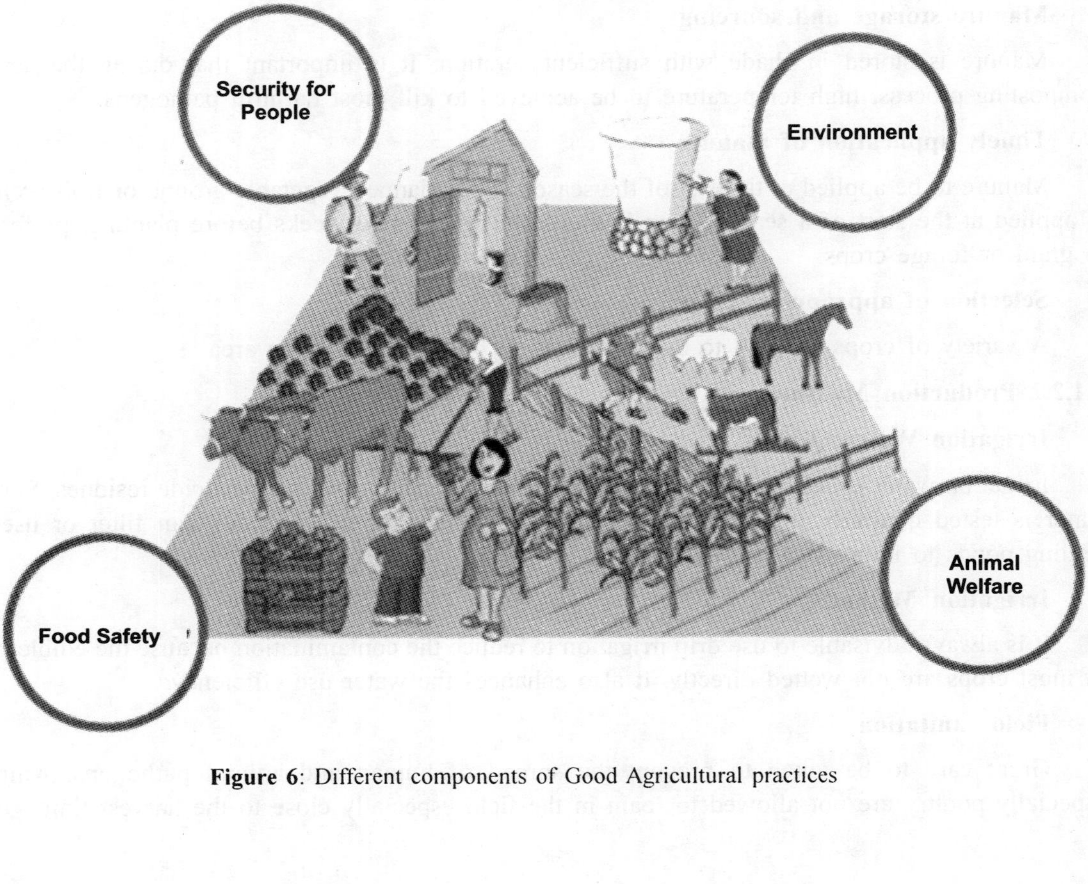

Figure 6: Different components of Good Agricultural practices

Now GAP is formally recognized in the international regulatory framework for reducing risks associated with the use of pesticides, taking into account public and occupational health, environmental and safety considerations. This increasing trend of popularity acceptance of GAP by the consumers and the retailers provides incentives to the farmers by paying a premium wherein farmer would find alternatives to reduce the contamination right from the sowing of crop to harvest. GAP applies to a wide range of food/agricultural commodities that include fruits and vegetables, dairy products, medicinal and aromatic herbs, ornamentals, aquaculture etc.

4.1.2 Risk Minimizing Measures

Some of the risks minimizing measures in GAP are mentioned below

4.1.2.1 Pre-planting Measures

Site Selection

Land or site for agricultural production selected is on the basis of land history, previous manure applications and crop rotation.

Manure handling and field application

Proper and thorough composting of manure, incorporating it into soil prior to planting and avoiding top dressing on plants are important steps to be followed.

Manure storage and sourcing

Manure is stored in shade with sufficient aeration. It is important that during the aerobic composting process, high temperature to be achieved to kill most harmful pathogens.

Timely application of manure

Manure to be applied at the end of the season to all planned vegetable ground or fruit orchard. If applied at the start of a season, then it should be spread two weeks before planting, preferably to grain or forage crops.

Selection of appropriate crop

A variety of crops adapted to the local area to be cultivated in an area.

4.1.2.2 Production Measures

Irrigation Water Quality

Irrigation water should be free from undesirable salts pathogens and pesticide residues. Surface water is tested quarterly in the laboratory for any contaminations. Farmers can filter or use the settling ponds to improve water quality.

Irrigation Methods

It is always advisable to use drip irrigation to reduce the contamination because the edible parts of most crops are not wetted directly. It also enhances the water use efficiency.

Field sanitation

Great care to be taken to prevent the spread of human and animal pathogens. Animals especially poultry are not allowed to roam in the field especially close to the harvest time.

Worker facilities and hygiene

The farm workers are provided with hygienic well maintained toilet facilities around the farming areas. Farmers should get proper training to make them understand the relationship between food safety and personal hygiene. These measures are to be monitored and enforced.

4.1.2.3 Harvest Measures

Clean harvest aids

Baskets, bins and all crop containers have to be washed and rinsed properly. They should be properly covered when not in use to avoid contamination by birds and animals.

Worker hygiene and training

Good personal hygiene is very important during the harvest of the crops. Employee awareness, meaningful training and accessible rest room facilities with hand wash stations encourage good hygiene.

4.1.2.4 Post-harvest Measures

Worker hygiene

Packaging area should be clean and sanitized. The worker should be clean and use disposable gloves on packing lines.

Monitor wash water quality

Potable water should be preferably used in all washing operations. Use chlorinated water to wash the fresh produce.

Sanitize packing house and packing operations

Loading, staging and all food contact surfaces should be cleaned and sanitized at the end of each day. Care taken to prevent rats and rodents, in the packing house. Packaging material, to be stored in a clean area.

Pre-cooling and cold storage

Harvested produce to be quickly cooled to minimize the growth of pathogens and maintain good quality. Refrigeration room should not be overloaded beyond cooling capacity.

Transportation of produce from farm to produce

Cleanliness of the transportation vehicles to be maintained. For traceability norms, it must be ensured that each package leaving the farm can be traced to field of origin and date of packing.

The above mentioned GAP can be followed in many parts of India if the farmers organize themselves as a consortia or federation. Like VFPCK in Kerala and TNSFAC in Tamilnadu are few success stories. It will not be possible for an individual small farmer to follow all the guidelines due to many limitations. Nonetheless, GAP ensures food safety from the farm to fork.

4.2 Standards and Regulations for Organic Products

Organic market growing by leaps and bounds warrants stringent regulation to assure quality and prevent fraud. This is amply illustrated by the variety of labels claiming organic in innovative ways like, natural, organically grown, chemical free, etc. However, for consumers who have no access to the farms where the produce is cultivated "certified organic" serves as a product assurance. On the other hand, certification is a marketing initiative aimed at regulating and facilitating the sale of organic products to consumers. Being able to put the word "organic" on a food product is a valuable marketing advantage in today's consumer market and makes life easier for the discerning consumer. Under the certification rules, a product made with fully organic ingredients can carry the mark" 100% organic". Products with 95% organic ingredients can use the word "organic" and those that contain a minimum of 70% organic ingredients can be labeled as "made with organic ingredients". In addition the products may carry the logo of the certification agency that has approved them. These rules are generally applicable in all nations. In general all businesses directly involved in organic crop/food production can be certified whether they are seed suppliers, farmers, food processors, traders or retailers. Standards are the expected level of attaining various components and levels of implementation, achievements, and adherence of organic farming as may be defined by the authorities, like FAO, IFOAM, European Union, USDA etc. The organic standards have been enunciated and enshrined in various documents such as Codex Alimentarius and IFOAM charter and adapted by various countries. The number of countries with organic standards has increased to 73, and there are 16 countries that are in the process of drafting legislation (Table 1). Together, the IFOAM Basic Standards for Organic Production and Processing (IBS) and the IFOAM Accreditation Criteria for Bodies Certifying Organic Production and Processing (IAC) constitute the IFOAM norms. These norms are the basis for IFOAM's organic guarantee system. This system unites the world through a common set of standards, verification and market identity. The accreditation program is a service offered to certification bodies.

Table 1: Legislations for organic produce across different countries in the world

Country/Union	Legislation	Handling agency
European Union	EU 2092/91	Soil Association/ Organic Farmers and Growers
Sweden	Jordbruksverket	KRAV
US	National Organic Program (NOP)	US department of Agriculture(USDA)/Quality Assurance International (QAI)
Japan	Japanese Agricultural Standard	Ecocert-QAI Japan Ltd
Canada	National organic standard	Private companies
Australia	National Association for Sustainable Agriculture, Australia (NASA)	Australian Quarantine and Inspection Service/ Biological Farmers Australia
Newzeeland	Food Standards Australia New Zealand (FSANZ)	New Zealand Food Safety Authority
Switzerland	Swiss organic farming ordinance	Bio-Suisse
Germany	Naturland	
India	National Organic Standards	National Programme on Organic Production

Contd...

Bhutan	National Organic Farming Framework (NFOFB)	Ministry of Agriculture and Forests / Department of Agriculture Planning and Policy Division /National Soil Service Centre/National Plant Protection Centre
China	China National Organic Product Standard	Standardization Administration of China
Egypt	EN45011 (ISO65)	Center of Organic Agriculture in Egypt
Israel	National standard for organically grown plants and their products	Plant Protection and Inspection Services
Cuba	Cuban Organic Agriculture Exchange Program	Organic Consumers Association
Korea	Eco-friendly Agriculture Promotion Act	Korean Food and Drug Administration (KFDA)
Taiwan	National Organic standard	Taiwan Council of Agriculture
Brazil	MAPA; Diario Oficial Regulation Act	Ministério da Agricultura, Pecuária e Abastecimento
Pakistan	Pakistan Pure Food Laws (PFL)	The Pakistan Standards and Quality Control Authority
Bangladesh	Bangladesh Pure Food Rules	Bangladesh Food Safety Authority
Finland	Luomuliitto	Finnish Food Safety Authority(EVIRA)

4.3 Organic Certification

With the organic market growing steadily, organic certification and regulation associated with it is becoming increasingly stringent and mandatory. Organic certification is a certification process for producers of organic food and other organic agricultural products. In general, any business directly involved in organic food production can be certified including seed suppliers, farmers, food processors, retailers and restaurants. Requirements vary from country to country and generally involve a set of production standards for growing, storage, processing, packaging and shipping that include:

i) avoidance of synthetic chemical inputs (*e.g.* fertilizers, pesticides, antibiotics, food additives, etc).

ii) use of farmland that has been free from chemicals for a number of years (often > 3 years).

iii) keeping detailed written production and sales records maintaining strict physical separation of organic products from the noncertified products.

iv) Undergoing periodic on-site inspections.

4.3.1 Purpose of Certification

Certification addresses a growing worldwide demand for organic food and is intended to assure quality and prevent fraud in organic trade. It helps all the stakeholders in the production and distribution chain. Certification helps the organic producer to identify himself as a supplier of products approved for use in the certified operations. While for the consumers certification serves as product assurance. In other words, certification is essentially a marketing initiative aimed at regulating and facilitating the sale of organic products to consumers. Individual certification bodies have their own logo which serves as brand to consumers. However, certification standards and organic laws do not affect the existing national policies or legislation.

4.3.2 Certification Process

Broadly the certification process is divided into two parts *i.e.*, inspection and certification.

Inspection aims to verify, the production and handling are carried out in accordance with the standards against which certification will be carried out. It is important for the farmer or farmer federation intending to go for certification to be acquainted with the following points:

- Understanding organic standards pertaining to crop production, storage, transportation and marketing.
- Compliance with the standards which may warrant modification of the existing practices.
- Documentation with reference to infrastructure, field history, resource management and record keeping system.
- Production planning covering all facets of activities.
- Inspection by the inspector coupled with oral interview and collection of samples(soil, water, plant, produce, input) for analysis.
- Keeping of record covering all activities to be made available for inspection, at any time of the year.

Certification aims to confirm the production and handling is in consonance with the standards. The certification committee members will take the certification decision based on the available documentation. In cases where decision is positive, the decision will be notified within three months after the inspection report and missing documents if any, has been received. Certification logo either 'certified organic' or 'in-conversion' as approved by the certification committee, will be sent to the farmer or the farmer federation as a soft copy. The farmer then makes the label incorporating the logo and other relevant details and sends it to the certification office for approval. The label approval committee of certification office approves the label as per the standards and sends the approval decision along with certificate to the farmer and only then the farmer or producer can use the label. If the producer has valid reasons not to accept the certification decision, the concerned party can request for reconsideration of the decision.

Certification is important for the following reasons

- Provides for product differentiation.
- Ensures product's value.

- Protects consumers from fraud.
- Boosts consumer confidence.
- The National Organic Program requires certification of agricultural products making an organic label claim.
- Assures organic integrity from seed to shelf.

In India, Agricultural Processed Foods Export Development Authority (APEDA) under Ministry of Commerce is the controlling body for organic certification for export. Till date there are no domestic standards for organic produce within India. Currently 11 certification agencies have been authorized to undertake certification process under National Programme for Organic Production (NPOP).Around the world different organic logos are used (Figure 7).In countries without organic laws, government guidelines may or may not exist, while certification is handled by non-profit organizations and private companies.

Figure 7: Labels of international organic certification bodies across the countries

4.4 Documentation of Farm Records and Traceability

Documentation of all the production processing and marketing activities with complete details to facilitate traceability at any point of time is an important component of inspection and certification.

<table>
<tr><th colspan="2">Documentation</th></tr>
<tr><td colspan="2">Internal Control System
(documented quality assurance system for external certification body)
Developing Internal Control System</td></tr>
<tr><td>The operator and its policy</td><td>Description of production and processing</td></tr>
<tr><td>1. Organization (manpower and experts)</td><td>• General Project Area (terrain, altitude, temperature, rainfall).</td></tr>
<tr><td>2. Activities</td><td>• Methodology and materials (resource availability, package of practices, resource management method)
• Buying, handling and exports (database of staffs involved)
• Risk assessment (contamination/mixing)
• Organization (flow chart of hierarchies)</td></tr>
<tr><td colspan="2">Functions</td></tr>
<tr><td colspan="2">Task/Responsibilities</td></tr>
<tr><td colspan="2">1. Organic Approval Committee for organic growers, internal inspectors, project staff and organic produce
2. Training to growers and project staff
3. Registration of grower, NGO, Society or farmer group
4. Inspections
5. Yield estimation
6. Documentation of buying/selling
7. Storage and handling
8. Penalties for violation
9. Certification</td></tr>
</table>

CHAPTER 5

Nutrient Management in Organic Agriculture

5. Nutrient Management in Organic Agriculture

In organic agriculture 'feeding the soil' is more important than feeding the plant. Maintaining high humus content in the soil with appropriate soil management techniques is key to the success of organic agriculture. Some of the simple techniques that are very effective are mulching, green manures, cover crops, composting which can be implemented by the small holders by utilizing the local resources.

Why Soil organic matter (SOM) matters?

Biological functions of SOM

- provides nutrients and habitat for organisms living in the soil.
- provides energy for biological processes.
- contributes to soil resilience (the ability of soil to return to its initial state after a disturbance, for example after tillage).

Chemical functions of SOM

- measure of nutrient retention capacity.
- provides resilience against pH change.
- main store of many key nutrients especially nitrogen and potassium.

Physical functions of SOM

- binds soil particles into aggregates improving soil structural stability.
- enhances water holding capacity of soil.
- moderates changes in soil temperature.

A. Mulching

Mulching is a method of covering the soil with a thin layer of biomass. For mulching, leaves, bark, nut shells, weeds, grasses, wood chips, silage, paper, pine and conifer needles, paddy or wheat straw, rice husk, coir dust, saw dust, banana and sugarcane leaf trashes *etc* which are available in the region can be used. The benefits of mulching are to prevent the loss of water by evaporation and transpiration, keep down weeds due to soil solarisation, dampen temperature fluctuations, increase soil moisture storage and facilitate uniform distribution of moisture in the soil horizon (Figure 8). It reduces the runoff and soil losses, prevents crusting and soil compaction, and reduces blowing and beating action of water and wind.

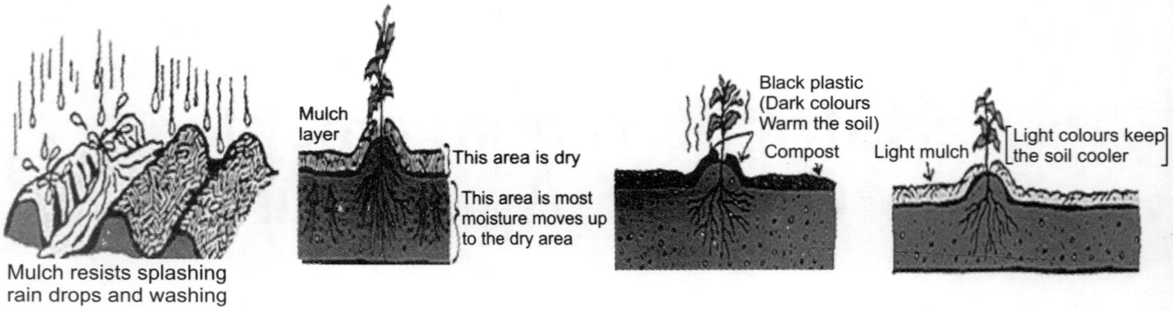

Figure 8: Different functions of mulching.

Mulches can be classified as inorganic or organic (Figure 9). Inorganic mulches include plastic, rocks, rock chips, and other nonplant materials. Plastic is the only inorganic mulch used in vegetable gardens. Organic mulches include straw, compost, newspaper, sawdust, and similar materials.

Mulches modify the micro-climate, alter the environment of soil microbes, enhance soil flora and fauna activity, modify soil moisture regimes and properties associated with it and soil temperature in the root zone, improve rooting environment and soil productivity. Water use efficiency can be doubled with mulching, frequency and intensity of irrigation can be reduced drastically. For mulching one hectare area of any agricultural crop about 4-5 tons of the biomass is required. In most of the agricultural crops, mulches should be applied uniformly after the first weeding which generally occurs after 2-3 weeks of sowing/transplanting. Mulching is also very useful in horticultural crops as it reduces the water requirement by reducing the evapo-transpiration. In most of the perennial crops like apple, pears, plums, mandarins, walnuts and fruit crops fallen leaves of the same plant can be used as mulch.

Figure 9: Different mulching options.

B. Green Manures

Green manures, also referred to as fertility building crops, may be broadly defined as crops grown for the benefit of the soil. They have been used in traditional agriculture for thousands of years but conventional farming systems largely rejected them as the use of fertilizers and pesticides became more common. Although they have many roles they are still often under utilised by today's organic farmers. However, recent emphasis on reducing the environmental impact of all farming systems (stimulated by new legislation) has led to a growing interest from the conventional sector.

Green manuring can bring a number of advantages to the grower:

- Adding organic matter to the soil
- Increasing biological activity
- Improving soil structure
- Reduction of erosion
- Increasing the supply of nutrients available to plants (particularly by adding nitrogen to the system by fixation)
- Reducing leaching losses
- Suppressing weeds
- Reducing pest and disease problems
- Providing supplementary animal forage
- Drying and warming the soil

A number of disadvantages can also be identified:

- Direct costs of seed and extra cultivations
- Lost opportunities for cash cropping
- Extra work at busy times of the year
- Exacerbated pest and disease problems (due to the 'green bridge' effect)
- Potential for the green manures to become weeds in their own right

A wide range of plant species can be used as green manures. Different ones bring different benefits and the final choice is influenced by many considerations. If the most is to be made from green manure crops it is important that they are carefully integrated into the crop rotation and proper attention paid to their husbandry.

Farmers with small holdings and who cannot afford to lose a season for the cultivation of green manure crops can undertake green leaf manuring. It is a process wherein the green leaves, tender green twigs collected from shrubs and trees grown on bunds, waste lands and nearby forest are ploughed into the soil. The common shrubs and trees used are glyricidia (*Glyricidia maculata*), subabool (*Leucania leucocephalla*), *Acacia sp., etc.* that are locally available in the region (Figure 10).

Figure 10: Different green manuring crops and green leaf manuring in field.

Multistoried cropping

The necessity to intensify agricultural land use has led to the development of multistorey cropping .The high density multi species mostly multipurpose plants cropping system utilizing sunshine at various vegetation layers (Figure 11) is an efficient system in energy utilization and building soil fertility.

Figure 11: Multistoried cropping.

Biofencing

Biofencing is the practice of establishing and maintaining perennial herbs, shrubs and trees that are resistant to biotic and abiotic stresses along the field boundaries (Figure 12). The clippings of the fence can be used for mulching and composting. Also it helps to prevent the soil erosion from the farm.

Figure 12: Biofencing.

Concentrated organic manures

Concentrated organic manures are the byproducts of animal or plant industry which are organic in nature and contain higher percentage of nutrients compared to bulky organic manures. The commonly available concentrated manures are oil cakes, meal group of manures etc.

Oil cakes

The solid portion left after the extraction of oil from oilseeds(Figure 13) is dried and used as manure. Both the edible and non edible oil cakes can be used as manure. The edible oil cakes like groundnut cake while the non edible oil cakes like castor cake are well powdered and used as manures.

Figure 13: Cotton, mustard and coconut oil cake.

Meal group of manures

Examples for the meal group of manures are bone meal, hornmeal and hoof meal (Figure 14). The fish mea prepared from non-edible and unsold fish is also used as manure.The use of meal is restricted as per the organic standards.

Figure 14: Meal group of manures.

C. Bio-fertilizers

Biofertilizers, more commonly known as microbial inoculants, are artificially multiplied cultures of certain soil organisms that can improve soil fertility and crop productivity. These preparations containing efficient strain of nitrogen fixing, phosphate solubilising or cellulolytic micro-organism used for inoculating seed or soil with the object to accelerate microbial processes to enhance availability of nutrients in the available form.

Bacterial Biofertilizers

Bacterial biofertilizer are the products containing cell of different types of beneficial bacteria (Figure 15). They are widely used in organic farming and some of the biofertilizers commonly used are furnished below.

a) Rhizobium

Rhizobium inoculants help in establishing efficient symbiotic association with pulses and leguminous fodder crops and can fix 50-100 kg N/ha and also leave sizeable amount of nitrogen to the succeeding crop in rotation. These inoculants are specific for each leguminous crop which is referred as cross inoculation groups. Therefore the inoculants should be used only for the recommended crop. These inoculants are available in the form of charcoal based cultures in 200 gm, 500 gm and 1kg packages.

b) *Azospirillum*

Azospirillum, a free living nitrogen fixer and occurs in association with roots of almost all plants of agricultural importance and can fix 20-40 kg N/ha. The crops which respond well to Azospirllum cultures are bajra, barley, oats, sorghum, millets like finger millet, Italian millet, kodo millet, barn yard millet, forage grasses and fodder crops. It is advocated as a seed treatment.

c) *Azotobacter*

Azotobacter inoculants are popular in India because it can be applied to a number of non-leguminous crops and promotes seed germination and initial vigour of plants due to growth substances produced by these organisms. It also suppresses the growth of some saprophytic and pathogenic microorganisms near the root system of crop plants and fixes 20-40 kg N/ha.

d) Phosphobacteria

Phosphorus is one of the important plant nutrients and is found in an unavailable form in soil. The introduction of phosphobacteria in the seed or soil increases the availability of phosphorus from the insoluble sources of phosphate in soil. Several soil bacteria, particularly those belonging to

Bacillus polymyxa, Pseudomonas striata and fungi belonging to the genus *Penicillium sp.* and *Aspergillus sp* possess the ability to transform the insoluble P to soluble forms by secreting organic acids. Phosphate solubilising microorganisms are recommended for all crops and can solubilize 20-30% of insoluble phosphate in the soil.

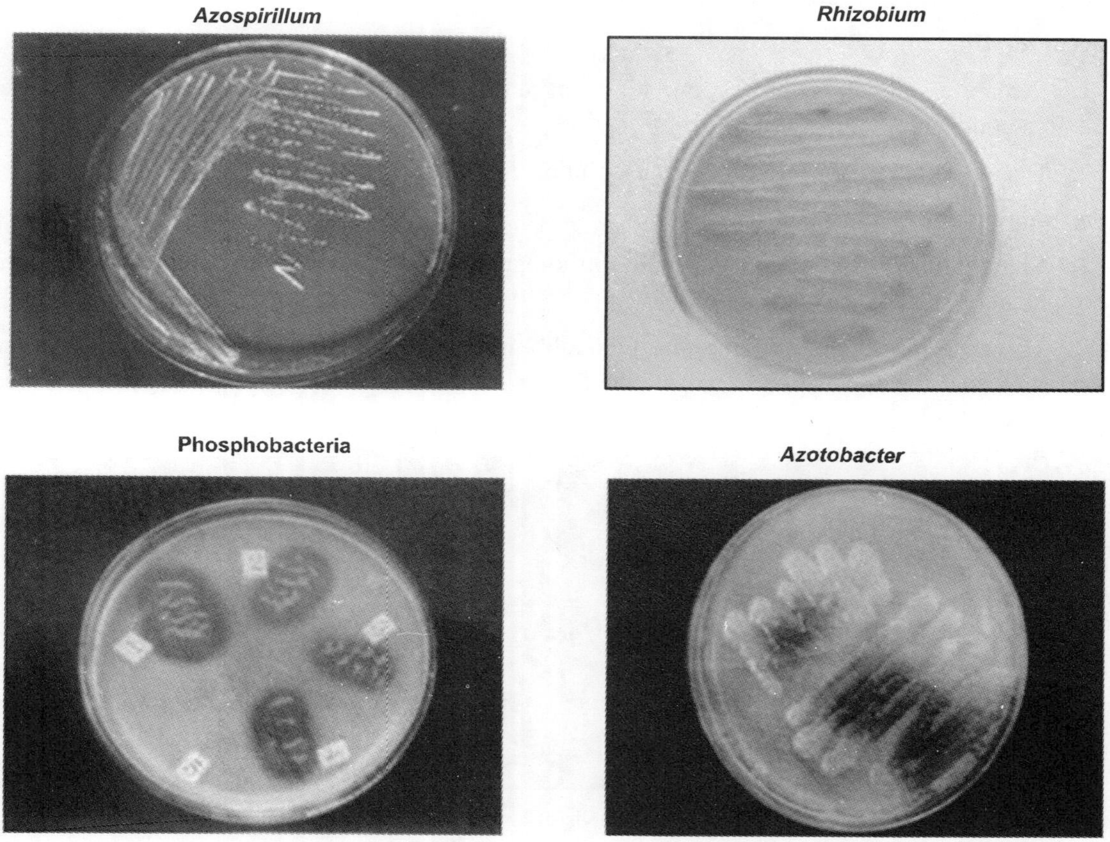

Figure 15: Bacterial biofertiliser.

Method of application of bio-fertilizers

There are three ways (Figure 16) of using these biofertilizers which have been explained below.

a) Seed Treatment

1. Prepare 5% jaggery solution by boiling 500 gm of jaggery in 10 litres of water for 15-20 minutes. Depending on the quantity of seeds to be treated sufficient solution has to be prepared.
2. The solution has to be cooled.
3. Mix the bio-fertilizer in the above solution. The general recommendation is one kilogram of bio-fertilizer is sufficient to treat seeds required for one acre.

4. Heap the seeds to be treated on a polythene sheet.
5. Pour the bio-fertilizer solution over the seeds and mix the seeds so that every seed gets a uniform coating on the surface.
6. Dry the seeds in shade and sown immediately.

b) Seedling Dip

1. Prepare a solution by mixing one kilogram of bio-fertilizer in 50 litres of water in a barrel or container.
2. Dip the roots of the seedling in the solution for 30 minutes before planting.

c) Soil application

1. Mix 5 kilograms of bio-fertilizer in 100 kilograms of well rotten cattle dung or farm yard manure (FYM).
2. This mixture is spread in one acre preferably in the evening hours. It is ideal to plough the land after the application or irrigate the field which facilitates the microbes to get established in the field.

Figure 16: Different methods of application of biofertiliser.

Precautions

1. Use the bio-fertilizers before the expiry date.
2. Protect the bio-fertilizer packet from direct sunlight.
3. The treated seeds should be dried in shade and sown as soon as possible.

E. Blue green Algae (BGA)

BGA, also known as Cyanobacteria, are a group of photosynthetic bacteria that many people refer to as "pond scum." BGA are most often blue-green in color, but can also be blue, green, reddish-purple, or brown. Blue-green algae generally grow in lakes, ponds, and slow-moving streams when the water is warm and enriched with nutrients like phosphorus or nitrogen. BGA contribute substantially to the fertility of the soil especially under tropical rice field conditions. In addition to contributing up to 30kg N/ha/season, these algae benefit the crop plants by providing a part of biologically fixed nitrogen and secreting growth promoting substances like vitamins and amino acids.Also they add appreciable amount of organic matter into the soil, solubilise insoluble phosphate and improve the physical and chemical nature of the soil. The BGA strains used in India are *Anabena spp., Nostoc spp, Aulosira spp* and *Tolypothrix spp* (Figure 17).

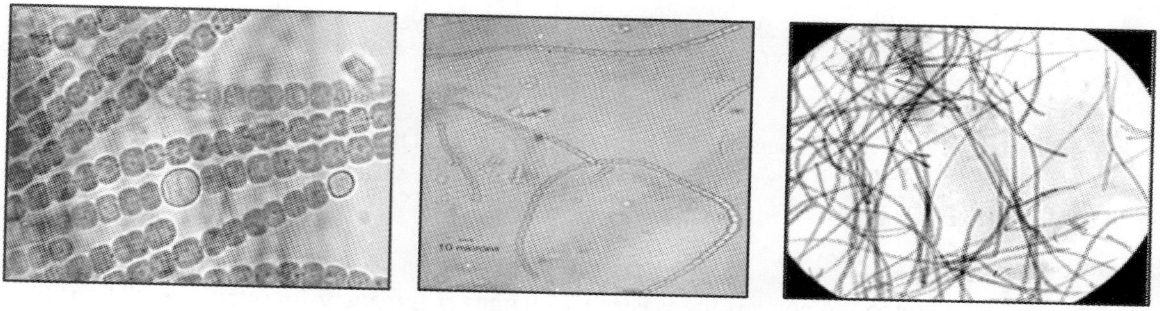

Figure 17: Common BGA strains.

Method of preparation

The algal inoculum can be prepared by the farmers at their fields by using rural oriented low cost production technology without any appreciable investment.

Trough method

i) Prepare a shallow tray (2 mt × 1 mt × 20 cm) of galvanized iron sheet (Figure 18) or cement or polythene lined pit on ground. The size can be increased if more material is to be produced.

ii) Spread 5-6 kg of sieved field soil which is free from any organic material. Fill the trays with water up to 5-10 cm and mix well with the help of long handle brush so that a uniform layer of soil is formed at the bottom. Spread 300-400 gm of rock phosphate.

iii) Sprinkle the starter culture of algae on the surface of the standing water, keeping trays in the open air and completely exposed to sun.

Figure 18: Trough method for algal inoculums production.

iv) In hot summer months, the growth of the algae will be rapid and in about 7-10 days a thick algal mat will be formed on the surface of water. The water level is maintained at 5-7 cm by adding water intermittently, when the growth of algae becomes very thick watering has to be stopped.

v) Allow the water to evaporate completely in the sun and dry algae cracks into flakes. The dry algal flakes can be stored in the polythene bag and used in the fields when required.

vi) Fill the trough with water and add very small amount of the dry algal flakes as inoculum without adding fresh soil. After 10-15 days, harvest the algal mat and repeat the process by adding fresh soil and other required material as mentioned above.

Field application of BGA

Broadcast 10-12 kg of soil based algal bio-fertilizer per hectare after 3-4 days of transplanting of rice seedlings in standing water. It should be ensured that the water level should be maintained for 7-10 days after inoculation.

F. Composting

Composting is method to convert the wastes into humus rich manure for improving the fertility and productivity of soil. It's very simple and apt for the small holder farmers. During the process of composting the biodegradable wastes like crop residues, weeds, leaf litter, animal wastes etc are scientifically transformed into manure which is called as compost. The compost is rich in beneficial soil microorganisms, provides the nutrition to the crops and improves the structure and texture of the soil.

Compost provides many benefits as a soil amendment and source of organic matter by improving soil biological, chemical, and physical characteristics viz.,

- Increases microbial activity.
- Enhances plant disease suppression.
- Increases soil fertility.
- Increases cation exchange capacity.
- Improves soil structure in clayey soils.
- Improves water retention in sandy soils.
- Reduces bioavailability of heavy metals.

There are different methods of composting which are practiced across the world, the simple and efficient methods are:

a) Heap method

b) Aerobic composting

c) Vermicomposting

G. Important Points to consider before composting

1. Sources of materials used for composting

The available biodegradable wastes in a region can be efficiently used for composting like:

i) *Crop Residues:* Straw, maize stalks, etc.

ii) *Forest resources*: Leaf Litter.

iii) *Weeds:* Different weeds that are available in the farm.

iv) *Green Manures:* Sunhemp, Cluster Beans, Sesbania etc.

v) *Livestock Wastes:* Cattle and shed wastes comprising cow dung and urine.

2. Selection of the Site

Important criteria of selecting the site for composting are as follows:

i) Availability of materials to be used.

ii) Collection and transportation of the material.

iii) Sources of water supply.

iv) Transportation of the finished compost to the field.

v) Distance of the field from the site of composting.

vi) Away from water logging and trees

Site selected for composting should be demarked permanently as it would build up desirable micro-organisms in the soil beneath and hastens the ongoing composting process.

1. Choice of structure

Generally there are two types through which wastes can be decomposed at the farm level. Mainly composting is done in a heap or tank like structure with provisions for aeration. Any one of these can be used. Heap method is economical and easy to manage.

2. Preparation of the material

Although different materials can be used directly, some may need pretreatment before adding to heap, as given below:

i) Too wet material containing high moisture should be allowed to wither a little.

ii) Rough and coarse materials such as stalks of maize, cotton, millets etc. must be broken or chopped before use. The ideal length is around 5 cm. These materials can be broken by spreading them in the cattle shed and it will also help in collecting urine and dung properly.

iii) Woody materials such as sugarcane trash, tree bark and saw dust should be moistened before putting to the heap, preferably by soaking them in water for 7-10 days.

***vi*) Construction of the heap**

The dimensions of the heap or tank are; width 2m, height 1m and the length depends on the quantity of the material. At the base of the heap a layer of old branches or wooden logs or big stones should be laid as this would provide aeration and prevent water logging.

***vii)* Turning the heap**

The heap is turned so that the material from the edges where the temperature is lower is moved to the centre where it is warm, while centre of heap does not suffer from overheating. All the materials in the heap should be exposed to the heat in order to kill germs, weed seeds, insects and also to speed up the decomposition. Turning also improves aeration of the heap. The first turning should be done after 2-3 weeks of heaping, when temperature has started to drop after reaching a peak. The heap can be turned again after three weeks of first turning. If the heap is prepared properly, not made compactly, holes are made and proper ratios of residues having different C: N ratio, is maintained; only one turning of the heap is enough.

***viii*) Maturing of compost**

The time required for the heap to mature and become good compost, depends on the local climatic conditions and the materials used. If the weather is warm, the heap is moist, well aerated and good combination of materials is used for building the heap, the compost is ready in three months time. In colder or dry conditions and dominance of materials with high C: N ratio, heap usually takes 4-5 months to ripe. When the compost is ripe the material is converted from dark brown to grey colour, gives an earthy smell and is coarse. The volume of the mature heap is reduced to half of the original.

H. Heap Composting

The heap composting method (Figure 19) is a very simple and is quite popular.

Materials Required

Green biomass – 1 ton; Dry Biomass – 1 ton; Cow dung – 200 kg and Water.

Method

1. Demark an area with 2metres × 3metres dimension. Depending on the availability of the materials the length of the heap can be extended.
2. A layer of stones or wooden logs are placed within the demarked area as a basal layer which helps in providing aeration.
3. Dung slurry is prepared by mixing 25 kg of dung in 100 liters of water.
4. A layer of dry biomass is uniformly spread to the height of 30 cms and moistened with dung slurry.
5. On the top of the dry biomass layer, green biomass is spread and moistened with dung slurry.
6. Similarly, alternate layers of dry and green biomass are laid to the height of about 1-1.5 metres height. Each layer of biomass is thoroughly moistened with dung slurry.
7. Similarly, alternate layers of dry and green biomass are laid to the height of about 1-1.5 metres height. Each layer of biomass is thoroughly moistened with dung slurry.
8. When the heap has reached a desired height it is covered with soil or straw.
9. To prevent the loss of nutrients by volatilization, the heap is made in shade. In certain places wherein there is no natural shade, thatched roofs or shade nets are to be made available.
10. To hasten the process of composting the heap can be turned after a month.
11. Depending on the weather conditions the compost will be ready in 2-3 months. In warm weather the composting process is faster than in winter months.

Basal layer for aeration

First layer of waste material

Preparation of cow dung slurry

Building up of heap with dry and fresh material

Finishing up of the compost heap (1-1.5m)

Thatched roofing

Figure 19: Heap composting.(Courtsey: Dr.A.Thimmaiah, Thimphu, Bhutan)

I. Aerobic Composting

Aerobic composting (Figure 20) is a very unique method of converting the wastes into humus.

Materials required

The wastes can be categorised into two types:

1. **Nitrogenous:** The materials with high nitrogen content *e.g.* green plant materials, kitchen wastes, dung and animal manures etc.

2. **Carbonaceous:** These are the more stable materials having high carbon content *e.g.* straw, dried leaves and crop residues after harvest, dried seaweed etc.

Size of the heap

The size of the heap should be 5 metres length, 2 metres width, and 1-1.5 metres height. The length can be extended based on the availability of materials. Such extended heaps are called windrows.

Method

1. Clean the area by removal of grass and other weeds with a spade.
2. Mark the area 5 m lengthwise and 2 m width.
3. An inverted 'V' channel is made with the help of dried stacks/ stones/bricks at the centre of the marked area running lengthwise. This tunnel will facilitate aeration.
4. Spread all the carbonaceous materials above the inverted 'V' shaped tunnel within the demarked area (5 m length × 2 m width) to a height of 15-30 cm.
5. Cow dung slurry or biogas slurry is spread over the carbonaceous layer in a very thin layer.
6. Over the slurry, the nitrogenous materials are evenly spread to a height of 15-30 cm
7. Again a layer of cow dung slurry or biogas slurry is spread over the nitrogenous layer.
8. Likewise a homogenous blend of carbonaceous and nitrogenous layers with dung slurry in between is made up to a height of 1.5 metres. On an average
9. 40% carbonaceous and 60 % nitrogenous material are ideal.

Figure 20: Different steps in composting.

10. The compost can be enriched with lime, rock phosphate or any such natural materials depending on the soil status.
11. The compost heap (5 m × 2 m × 1.5 m) is plastered with the slurry made out of equal proportions of cow dung and soil.
12. After 2 months turning the compost heap would enhance the composting process. However, there is no need to turn the heap. The compost would be ready in 2-3 months.

J. Vermicomposting

Vermicomposting is the method of converting wastes into compost by use of earthworms. Good quality compost is produced in a short period depending upon on the number of earthworms. In other words about 4-5 kg of wastes can be composted by 1000 worms (approximately 1 kg) in a day. The commonly used earthworms like *Eudrillus sp, Perionyx sp, Eisenia sp* or any locally available (Figure 21) surface feeding (epigeic) earthworms can be collected from the nearby soil and then used in vermicomposting. Vermicomposting can be undertaken in a tank or on the soil surface. The tank can be constructed with bricks or stones depending on the availability of materials. The height of the tank should be 0.75 metres, width 1 metre and length can be 3-4 metres or even more. The base of the tank should have a slight slope directing towards two drainage holes.

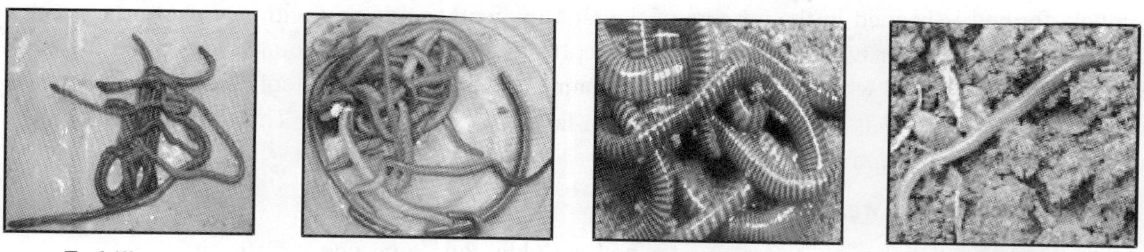

Eudrillus sp. ***Perionyx sp.*** ***Eisenia sp.*** **Locally available species**

Figure 21: Commonly used earthworms.

1. Basal Sand Layer

The first layer (5-10 cm) at the base is of sand as it helps in the drainage of excess water and also prevents the movement of worms into deeper layers of soil or out of the tank.

2. Dung Layer

On the top of the sand, a layer of one month old dung is laid. This layer is approximately 3-5 cm. thick.

3. Waste Layer

On the top of this dung layer all the biodegradable wastes are put and the tank is filled up to the brim (65 cms) with the waste.

4. Soil Layer

The wastes are covered with a thin layer (1-2 cm) of soil.

5. Dung Layer

Above the soil layer mature dung is uniformly spread (3-5cm) which forms the top layer in the tank. In other words, the wastes are sandwiched by two layers of mature dung (Figure 22). The contents are moistened with water (40-50% moisture level) for 7 days to predecompose the contents. After pre-decomposition, worms are introduced (1000 worms approximately 1 kg for 100-150 kg of wastes) and watered regularly to maintain 30-40 % moisture level for better activity of worms.

Watering is done in such a way the contents in the tank are not too soggy or too moist as it results in anaerobiosis of earthworms causing mortality. The tank is covered by a gunny sack in summer months to avoid evaporation of moisture. In places where the predators like rats, lizards, ants, pigs are a major problem; the tanks can be covered with wire mesh to prevent them feeding on the earth worms.

Collection of Vermicompost

Vermicompost appears on the top of the tank in the form of pellets. When this layer of vermicompost becomes a few centimeters thick watering should be avoided for 3-4 days. The reduced moisture at the top layer will force the worms to move down towards the lower layers. The compost formed is heaped in the corners of the tank by gently scrapping with a shovel or a wooden raft and left for a day or two. By this method of collection of wastes the young worms and the adult worms are not damaged while handling. In this manner the compost can be collected every 7-10 days till 80% of the material is collected *i.e* 10-15 cm layer is left at the base. The tank/heap should be refilled with pre-decomposed wastes.

Pre-decomposition of wastes

1. Spread a thin layer (10 cm) wastes and sprinkle cow dung and moisten it.
2. Likewise make layers of waste and cow dung up to 0.5 metres height.
3. Moisten the heap regularly with water for 10 days.
4. During the process high temperature develops and cools down in a week.
5. Check the temperature of the heap by inserting a stick.
6. If the stick heats up, leave the decomposing material for 3-4 days more and when the temperature is normal then only it should be used.

Refilling the Tank or Heap

Such pre-decomposed material is put on the heap or tank up to 75% of its height. The standard procedure of vermicomposting as mentioned above is continued i.e over the pre-decomposed waste, thin layer soil is spread and above it dung is spread. The worms come on the top and start feeding on the waste.

Preparation of Vermicompost tank

Waste layer on the dung & sand layer

Layer of fresh material

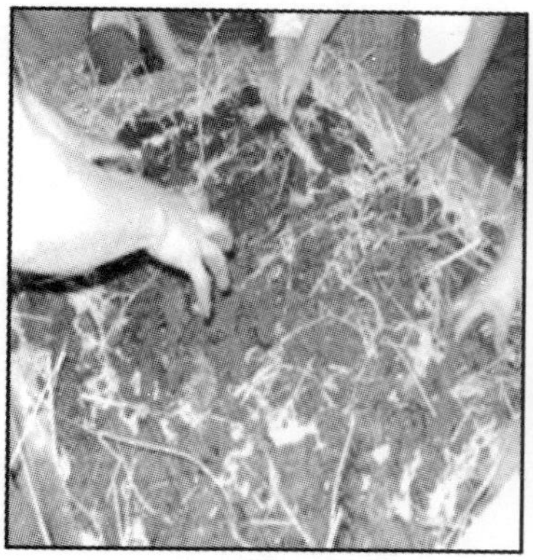

Final dung layer

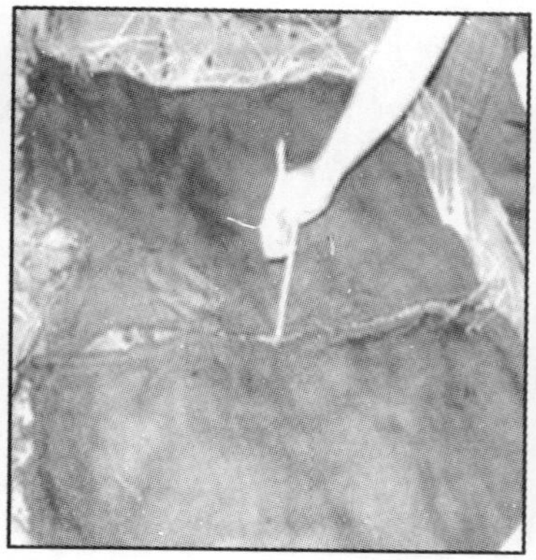

Gunny sack to maintain moisture

Turning gunny sack to test about the maturity of the product

Worms Separation

Final Product

Packed Product

Figure 22: Vermicompost production.

Precautions

1. Check the temperature of heap /tank by introducing a stick. If the stick heats up worms should not be introduced nor refilled.
2. The tank /heap should be moistened with water regularly.
3. Appropriate shade should be provided at the site where vermicomposting is undertaken.
4. Wherever the predators like rats, lizards, ants pigs etc., are found, wire mesh should be put on the top of the tank.

K. Characteristics of good compost

Following are the ideal compost compost characteristics:

- It should be dark brown to black with no particles of the original residue present.
- The organic matter content should be at least 80% (oven dried basis).
- The moisture content should range between 30-35%.
- The compost should have a Water Holding Capacity (WHC) of 150-200%.
- The organic carbon content should be 25.4-27.5%.
- The total nitrogen content should be 1.0 to 2.5%.
- The total P_2O_5 content should be 1.0 to 1.5%.
- The total K_2O content should be 1.0to 1.5%.
- The calcium content should be 4.2-6.7%.
- The magnesium content should be 0.2-0.3%.
- The sulphur content should be 0.4-0.5%.
- The cation exchange capacity (CEC) should range between 75 to 100 me/100 gm.
- The pH value should be between 5.5 and 6.5
- Odour should be slightly musty or earthy like forest soil.

L. Vermiwash production

Vermiwash is the coelomic fluid of earth worms. It contains anti fungal, anti bacterial and growth promoting substances. Vermiwash is a liquid that is collected after the passage of water through a column (Figure 23) of worm action and is very useful as a foliar spray.

Method

i) A plastic barrel with 200 litres capacity is used for making vermiwash. The tap should be attached at bottom/lower side of the barrel to collect the vermiwash.

ii) A sand layer of 2 inches thick is placed at the bottom. Over this layer a 2 inches thick small gravel layer is spread. On which two inch thick medium sized, and large sized gravel layers are placed respectively.

iii) Half feet thickness of one month old cow dung is spread over the gravel layer. Over this layer 2 feet thick slow degradable material (coconut fronts, pine needles) is spread. Over this layer pre decomposed waste material is spread up to 80 % volume of the barrel.

iv) One inch thick soil layer is laid over the layer of waste material. The soil layer is covered with ½ feet thick layer of old cow dung.

v) Facilities should be created for the water to trickle down the barrel at the rate of 20-25 litres/ day. Simple technologies like misters, pipe with tiny holes or even pot with hole etc. can be utilized.

Collection

i) The tap should be always kept open to collect the washings.

ii) The unit starts yielding good quality vermiwash after one week.

iii) The vermiwash is coppery brown in color.

iv) Vermiwash should be stored in cool dry place.

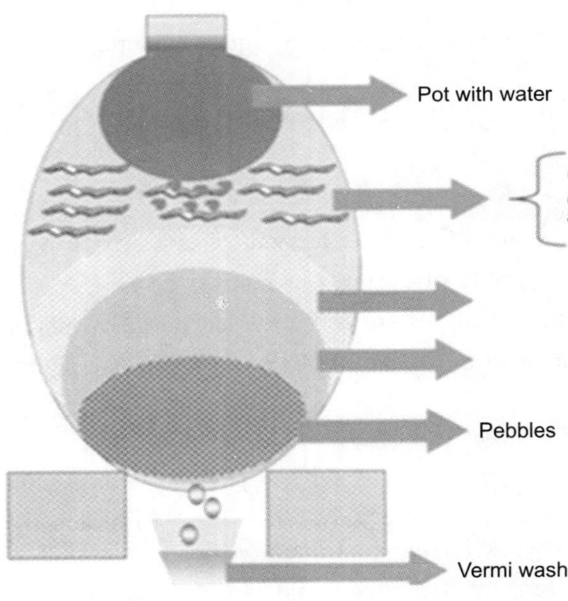

Figure 23: Vermi wash production.

Uses

Root dip/stem dip.

The seedlings before transplanting are dipped in vermiwash solution which is diluted 5 times with water for 15-20 minutes and then transplanted. Similarly the cuttings can also be dipped in the solution.

Foliar spray

Vermiwash is diluted in water 5 times and sprayed on the foliage of crops. It provides the plant with vital nutrients but also helps to control plant disease.

Soil drench

Vermiwash is diluted 10 times with water and the soil is drenched with the solution to prevent some of the soil borne pathogens.

M. Liquid manures

Liquid manures are the fermented preparations (Figure 24) obtained by active fermentation of plant or animal residues. It is prepared using cow dung and cow urine as base material and addition of green plants, leaves and tender stems of plants that are available locally like

i) *Green manuring plants*– Sunhemp, Dhaincha, Sesbania, Erythrina etc. and other leguminous plants can be used.

ii) *Tree leaves*– Leaves of other local medicinal trees.

iii) *Tender stems– Artemesia sp. Eupatorium sp.* and local plants.

iv) *Weeds*– Parthenium, Stinging nettle, and other weeds before flowering.

A mixture of different plants results in good quality liquid manure.

Materials Required

Cow dung-3kg; Cow urine-3lts; Green materials/Weeds-3kg and 20 litres capacity water bucket.

Method

The basic principle is to allow the materials to ferment over a certain period:

1. A plastic water bucket of capacity of 20 litres or more is taken.
2. The plant materials for liquid manures are chosen based on the availability.
3. The plant materials are shredded into small pieces and 3 kg of the plant material is put into the bucket.
4. Cattle dung 3 kg and cattle urine 3 litres are added to the bucket.
5. The bucket is filled with good quality of water and covered with a gunny sack.
6. The contents are stirred for minute's everyday for 7 days. To hasten the fermentation the contents are stirred once a week for the next three weeks.
7. The biomass will get fully fermented and the liquid manure would ready in a month for use.

Use

The liquid manure has to be diluted with water prior to application. The standard dilution is one part liquid manure in ten parts water and sprayed on the foliage.

Advantages

1. Act as tonic or stimulant to plants.
2. Is used as insecticide and fungicide when *Artemesia,* Stinging nettle and other aromatic plants are used as raw materials for liquid manures
3. Improves the nutrient assimilation.

Precautions

1. Stirring the liquid manure is very important.
2. The concentrated liquid manure should not be sprayed on the plant as it scorches the plants.
3. The liquid manure should be sieved before spraying otherwise the sprayer nozzle gets clogged.
4. Spraying should be done in the evening or cool sunshine hours.

Chopping of fresh material

Material (3kg) filled into the bucket

Cow dung and urine added

Content stirred everyday for 1 week

Filled with water & covered with gunny sack

Figure 24: Liquid manure preparation.(Courtsey: Dr.A.Thimmaiah, Thimphu, Bhutan)\

N. Dung Brew

It is a fermented solution of cow dung and provides nutrition to the crops.

Materials required

Cattle dung - 5 kg; Cattle urine – 15 litres and Plastic bucket of 20 litres capacity.

Method

1. Add 5 kg of cattle dung and 15 litres of cattle urine in a bucket and mix it well.
2. Cover the bucket with a gunny sack or any material that provides sufficient aeration.
3. Stir the materials every alternate day for 15 days.
4. The dung brew is ready for spray after 15 days and can be stored for 2 months

Cow dung

Cow urine

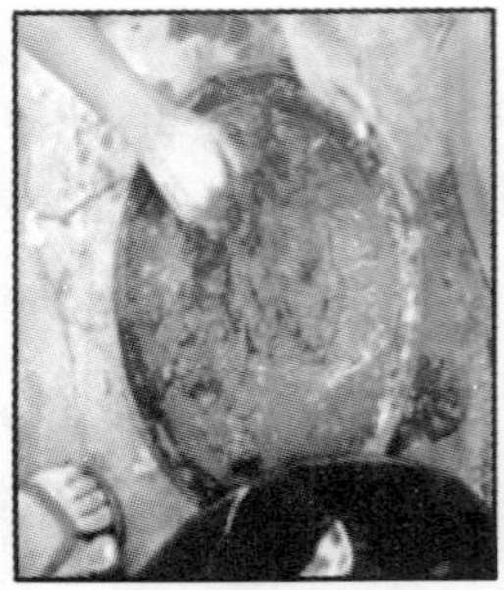

Mixing cow dung and Urine

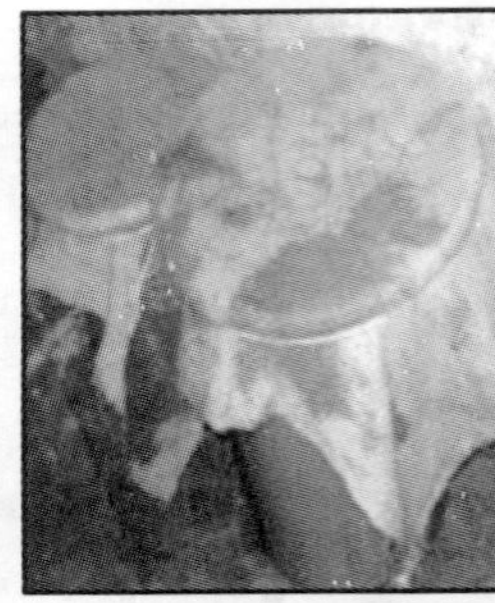

Covering with gunny sack

Figure 14: Preparation of dung brew

Use

One litre of dung brew is diluted in 10 litres of water and mixed thoroughly.The solution is sieved with a muslin cloth and then put into the knapsacksprayer and sprayed on the foliage of crops. Crops respond to this spray very well.

O. Compost Tea

It is a simple and effective preparation to enhance crop health. The efficacy of compost tea depends on the quality of compost prepared. Many farmers have observed that the compost tea is very effective against many pests and diseases.

Materials required

Well matured compost and water.

Method

1. Good quality compost has to be prepared by adopting heap, aerobic or vermicomposting methods.
2. One part of compost is immersed in 10 parts of water in a bucket or barrel depending on the quantity required.
3. The contents are stirred 5 minutes every day for 5-7 days.
4. The compost tea is ready to be used as a foliar spray after 5-7 days.
5. Before the application the solution is filtered and sprayed using a knapsack sprayer.

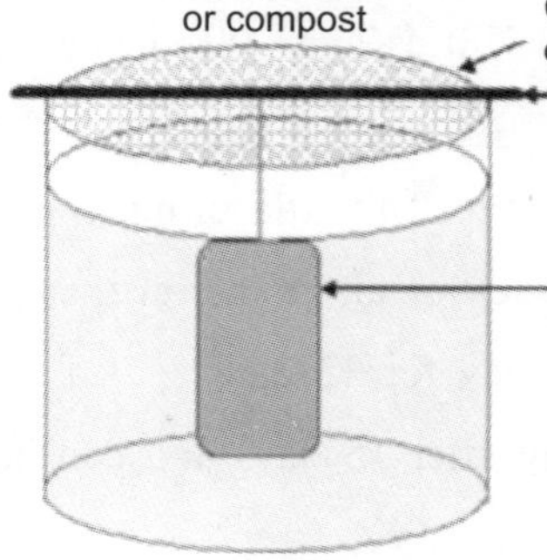

Figure 25: Compost tea production

P. Cow Urine spray

Cattle urine (Figure 26) is a good plant growth promoter which farmers can use on all crops. This resource is hardly utilized by farmers and is wasted in most of the cases. Cattle urine has more than 24 different nutrients that are vital for plant growth.

Materials required

Cattle urine and Water.

Method

Cattle urine is diluted 10 times with water and used as a foliar spray. Care has to be taken that the concentrated cattle urine is not sprayed on the crop as it will result in scorching effect.

Figure 26: Cow urine extract and its application.

Q. Stem Paste

It is an effective way to provide nutrition through the stem (cambium) of orchard crops. In addition to nutrition it prevents the attack of stem boring insects, mealy bugs and scales.

Materials required

Cattle dung – 50 kg; Sieved sand – 10 kg; Soil from the field – 10 kg and Cattle urine – 25 litres.

Method

1. Mix cattle dung, sand and soil with cattle urine to form a thick paste. Add sufficient cattle urine till the consistency is reached.
2. The stem of the tree is cleaned with a gunny cloth or brush with soft bristles.
3. The paste is plastered over the stem and branches of the tree.
4. The paste can be used as healing the cut end of the tree after pruning.

Figure 27: Preparation of stem paste and application.

R. Panchagavya

Panchagavya is an ancient Indian preparation which has been used for generations for human, animal and soil health. It consists of 5 products of cow viz., dung, urine, milk, curd (yoghurt) and ghee (clarified butter). When these ingredients are suitably mixed they have miraculous effects. However, to enhance the efficacy of Panchagavya, sugarcane juice, tender coconut water, ripe banana and toddy or grape juice is added depending on the availability of these materials. It acts as a general tonic to soil and crops.

Materials required

Fresh cow dung – 5 kg; Cow urine – 3 litres; Cow milk – 2 litres; Cow ghee (clarified butter) – 500 grams; Curd (yoghurt) made from cow's milk – 2 litres; Sugarcane juice – 3 litres; Tender coconut water – 3 litres; Ripe banana – 12 numbers; Palm/coconut or grape juice – 2 litres; Mud pot (wide mouth).

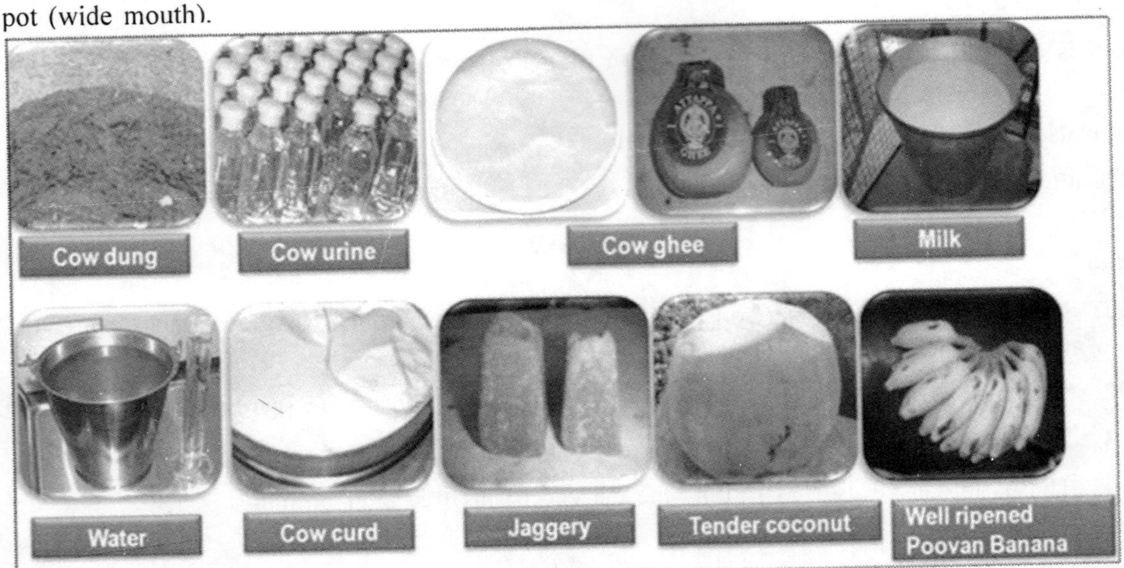

Figure 28: Ingredients of Panchagavya. (*Courtsey*: Nature Girl Solutions Pvt. Ltd.)

Method

1. Mix thoroughly cow dung and ghee in a container for 15 minutes three times a day for 3 days.
2. On the 4th day add the rest of the ingredients and stir the contents twice daily for 15 days.
3. The panchagavya preparation is ready after the 18th day.
4. The container should be kept in shade covered with wire mesh or a mosquito net to prevent the files from laying eggs.
5. If sugarcane juice is not available, add 500 grams of jaggery or palm sugar dissolved in 3 litres of water.
6. If toddy or grape juice is not available, add 100 grams of bakers yeast and 100 grams of jaggery in 2 litres of warm water. After 30 minutes add this solution.
7. Another method to make toddy is to keep 2 litres of coconut water in a closed container for 10 days, it gets fermented and turns into toddy and can be used.
8. When panchagavya is stirred regularly twice a day, it can be stored for 6 months without any deterioration of quality.

Usage

As foliar spray

For most of the crops 3% panchagavya is very effective *i.e.* 3 litres of panchagavya in 100 litres of water is ideal as a foliar spray. After dilution the panchagavya solution has to be filtered before spraying to avoid clogging of the spray nozzle.

Through irrigation

Panchagavya solution can be mixed with irrigation water at the rate 20 liters/acre either through surface irrigation or drip irrigation systems.

Seed/Seedling treatment

To drench the nursery bed, soaking the seeds or dipping the seedlings, 3% panchagavya solution is recommended. The rhizomes of tumeric, ginger etc can also be soaked in panchagavya. The soaking time for the planting materials is 20-30 minutes.

S. Biodigester

Biodigesters are water proof containers made using bricks and cement to ferment the biomass that are available in the farm. Making liquid manure from weeds and other plants available in the field provides a cheap and effective way of increasing and maintaining garden fertility and capturing minerals from deep within your soil.

Generally, the requirements for making the liquid manure are green plants (weeds, plants that have no use etc), animal manure and cattle urine.

Method

An effective brew is made from green plants. To make this, collect a variety of green plants. These can be prunings (strip the leaves from branches) or weeds pulled from the garden. A combination of tree leaves, grasses and soft and hard weeds will result in a good mix of nutrients. Weeds from the garden will contain stored plant nutrients. In fact, these weeds may reduce the leaching of valuable goodness from the soil. The pulled weeds can be left on the surface as mulch but put through a liquid fertilizer cycle will maximize their effectiveness and kill any seeds. Trees, being relatively deep-rooted compared with flowers and vegetables tend to uptake a different combination of minerals. Composting, mulching or making liquid manure from these will enrich the garden. These materials are filled upto the brim of the tank. A layer of cattle manure of 15 cm height is spread on the wastes (Figure 29). If cattle urine is available in plenty the tank can be filled with urine otherwise potable water can be used. Care should be taken the tank is covered with a thatched roof to prevent the rain and sunshine on the degrading material. The degradation of the material will be completed within 30-45 days. The outlet of the tank at the base can be opened to collect the solution. Once part of the solution is diluted in 10 parts of water and sprayed on the foliage of the crops.

Figure 29: Design of biodigester.
(*Courtsey*: Dr. A. Thimmaiah, Thimphu, Bhutan)

Benefits of Biodigester

The solution that is produced after the fermentation in a biodigester can be used for the following purposes:

1. To provide nutrients to crops by foliar sprays.
2. To protect crops from pests and diseases.
3. Help to avoid the use of synthetic chemicals like fertilizers and pesticides.
4. The local resources can be used efficiently.
5. Helps to save costs in crop production.
6. The preparations are environmentally friendly and protects the beneficial organisms.
7. Helps to use the local and traditional knowledge.
8. Empowers the farmers to be self reliant.

CHAPTER 6

Biotic Stress Management in Organic Agriculture

Organic Pest and Disease Management

In organic agriculture the pests and diseases can be managed by a cultural practices and biological approaches. These practices are region and crop specific.

A. Cultural practices

The cultural methods are the traditional practices followed by farmers for modulating crop growth through adjustments in time of sowing, cultural practices and plant population density for altering the microclimate to minimize the risks.

a) Crop rotation

Rotating the crop belonging to one family with one of a different family helps to solve the pest to a large extent (Figure 30). Rotating groundnut with maize will reduce the attack of white grubs. Rotating pigeon pea or chickpea with other non leguminous crop is a good practice for managing fusarium wilt and nematode problems.

Figure 30: Crop rotation.

b) Trap crops

A trap crop is a crop that is planted to lure insect pests away from the cash crop. The trap crop can be a different plant species, a different variety, or just a different growth stage of the same species, as long as it is more attractive to the pests when they are present. Successful use of trap crops is challenging. The trap crop must be more attractive to the pest than the cash crop, and steps must be taken to ensure that the pests in the trap crop do not later migrate to the cash crop. The pests in the trap crop can be eliminated by crop destruction .Mustard is a trap crop along with cabbage for the control of diamond back moth, aphid and leaf webber. African marigold is a good trap crop for the american boll worm, besides it also attracts the adults of leaf miner for egg laying on the leaves. Maize plants can be a trap crop to attract fruit fly adults in vegetable cultivation wherein fruitfly is a major problem and also for cotton bollworm.

Figure 31: Trap crops.

c) Intercropping

Intercropping generally has positive effects in terms of reducing the occurrence of insect pests. Insects find difficult to locate host plants as the visual and chemical stimuli for the host are not so strong and aromatic odour of other plants can disrupt host finding behaviour. Intercropping also interferes with the population development and survival of insect pests. For example, cabbage along with carrot or tomato is an important intercrop combination to effectively manage diamond back moth. Sowing cowpea as intercrop with groundnut minimizes leaf miner infestation. Green gram intercropped with sugarcane reduces the incidence of sugarcane early shoot borer. Growing short-duration pulses like black gram, cowpea, soybean and green gram as intercrops in cotton increase the effectiveness of natural enemies like coccinellids, syrphids, trichogrammatids *etc.*

Figure 32: Different intercropping options.

d) Use of resistant /tolerant varieties

Genotypes showing tolerance and resistance to insect pests and diseases are to be used for sowing. Plants have their sophisticated mechanism to protect them from the attack by pests. Certain genotypes act as detterents, antifeedants and some encourage the predators of pests. There should be constant watch to update such genotypes in the region. A series of resistant varieties are available for all the crops in accordance to the agro-climatic zone.

e) Traps

Insect traps are successfully used to capture the pests. Light traps attract night flying insects like moths, midges and some beetles. It consist of a source of light, such as electric bulb (200 watts), baffle plates and a collecting container. The most efficient light traps are fi tted with mercury vapour lamps. Sticky traps are useful for capturing insects which are active during both day and night. A sticky trap consists of a cylinder covered with a sticky material like vegetable oil. This is set up on a stick at crop level or at a suitable height. The insects get trapped and get killed. Yellow sticky traps are very effective in capturing aphids, thrips and whitefl ies. The traps to a large extent are useful for monitoring of pest infestation which helps to take appropriate remedial action.

Figure 33: Different traps.

f) Spike thrust method

This method is very useful for boring insects in trees. These insects make a hole on the trunk and starts feeding inside the tree. Insects are killed by inserting a long needle or a wire into the hole. This is very effective method to control stem borers in tree crops.

g) Trenching

Some insects like armyworms, red hairy caterpillars, nymphs of grasshoppers can be effectively checked by digging trenches around the field. The marching insects fall in this and can easily be killed.

Figure 34: Trenching.

h) Use of small flags and coconut fronds

These flags are made of plastic or paper sheets and strings attached to a long rope or plastic twine are examples of materials commonly used by local farmers to prevent birds from feeding on mature rice grains. These small flags are placed across the rice fields before harvest. Figure 35 presents inverted coconut fronds. Coconut fronds are placed in strategic locations in the rice field in an upside down position resembling an owl or a cobra to prevent rodent infestation.

Figure 35: Use of small flags and coconut fronds.

B. Different formulations for pest management

Following are the different formulations that can be used by the farmers in managing the pest by utilizing the locally available resources.

a. Liquid manure for pest management

A variety of plants (weeds) which have pesticidal value can be used. Plants which have strong disagreeable odour are ideal for making this preparation.

Materials required

Plants like *Parthenium, Lantana, Vitex, Eupatorium, Artemesia,* Stinging nettle etc., - 3 kg; Cattle dung - 3 kg; 20 litre capacity plastic bucket and Water – 20 litres.

Plant with pesticidal property: ***Horse tail, castor bean, Neem,Marigold***

Plants chopped into pieces

Chopped parts mixed with cow dung

Bucket filled with water and covered

Figure 36: Preparation of Liquid manure for pest control.

Method

1. Collect 3 kg of plants (leaves and tender parts) of plants which have pesticidal activity. Chop them into small pieces and put into 20 litre bucket.
2. Add 3 kg of cattle dung into the barrel and fill it up with water.
3. The barrel is stirred every day for 7 days and then stirred once a week for the next 3 weeks. The preparation will be ready in 30 days.
4. The concentrated solution is diluted ten times in water and used as a foliar spray.
5. These sprays are very efficient in managing a variety of pests.

Precautions

1. The liquid manure has to be diluted ten times before spraying on the crop otherwise it scorches the plant.
2. The solution has to be sieved through a cloth or gunny bag before spraying to avoid blockage of nozzles.
3. The efficacy of the solution is up to one month and has to be used within that period.

b. Organic Pest Repellent

The organic pest repellent can be made by the locally available materials. It is quite effective on vegetables and fruit trees.

Materials required

Artemesia leaves or Stinging nettle leaves or Lavender leaves – 200 gm; Hot chillies- 100 gm; Garlic – 100 gm; Egg yolk – 1 (acts as an emulsifier) and Any cooking oil – 100ml.

Procedure

1. Mix the above materials in a container or put them directly in a mixer grinder. These materials are grinded until the contents are turned into a fine paste.
2. Add 5 litres of water to the thick paste and mix thoroughly. This can be used as stock solution.
3. Take one part of the above concentration solution and dilute with 10 times with water. Stir the solution for 5 minutes and it is ready for spraying.
4. Strain the solution with the help of a muslin cloth before pouring the solution into a sprayer to avoid the clogging of the nozzles.
5. Spraying should be undertaken during evening hours.

c. Marigold Extract

Marigold (*Tagetes sp*) is an excellent pest repellent for most of the pests. When these plants are planted along with the main crop especially in vegetables, due to the allelopathic effect, they help in managing soil borne insects, pathogens and nematodes.

Materials required

Marigold flowers along with the entire plant and Soapnut or locally made soap.

Method

1. Take 5 kg of marigold plants along with the flowers. Grind into fine paste.
2. Add 100 litres of water and allow it for fermentation for 7 days.
3. Add 100 grams of soap nut powder on the day of spraying which acts as emulsifier.
4. Stir for 15 minutes and spray during evening hours.

Figure 37: Marigold extract.

d. Garlic Brew

Garlic spray helps to manage a broad spectrum of pests.

Materials required

Garlic-200 gm; Cattle urine – 20 litres and 20 litre capacity bucket.

Method

1. Crush or grind the garlic into a paste and put into the bucket containing cattle urine.
2. Stir the materials for 5- 10 min and cover with a gunny cloth or cotton cloth which can provide sufficient aeration.
3. Allow the materials to ferment for 5-7 days.

Figure 38: Garlic brew preparation. (Courtsey: Dr.A.Thimmaiah, Thimphu, Bhutan)

Method

1. Crush or grind the garlic into a paste and put into the bucket containing cattle urine.
2. Stir the materials for 5- 10 min and cover with a gunny cloth or cotton cloth which can provide sufficient aeration.
3. Allow the materials to ferment for 5-7 days.
4. After 7 days sieve the solution before spraying to avoid clogging of the nozzles of the sprayer.
5. The solution is diluted 10 times with water and sprayed on the foliage in the evening hours.
6. Care should be taken that the concentrated solution should not be sprayed in the crop.

e. Neem Oil Spray

Neem oil is a good insect repellent and anti-feedant. It is effective on a variety of crops.

Materials required

Neem oil – 3 litres; Soap nut – 200 gm or local soap – 100 gm and Water – 100 litres.

Method

1. Mix 3 litres of neem oil in 100 litres of water and stir well.
2. Add 100 gm of local soap into the solution.
3. If soap nut is available boil the soapnut (200 gm) in 1 litre water and for 15 minutes. Squeeze the boiled soapnut in the same boiled water and use the solution as an emulsifier.
4. Spray the solution in the evening hours.

Figure 39: Neem oil and its spray in the field.

f. Curd or Yogurt spray

This spray is very effective for most of the crop diseases like blight, mildews etc.

Materials required

Curd or Yogurt and Water.

Method

1. Allow the curd to ferment for 3-5 days until one can sense a strong smell of alcohol.
2. Take one litre of curd and dilute in 30 litres of water. Stir the solution for 15 minutes and sieve the solution with a muslin cloth.
3. Spray the solution during the evening hours.

g. Cattle Urine Spray

Cattle urine is generally not collected by the farmers and to a large extent they are wasted. There a need to build capacities of the farmers about the multiple benefits of cattle urine in crop production and protection in agriculture.

Materials required

Cattle urine – 1 litre and Water – 10 litres.

Method

1. Dilute one litre of cow urine in 10 litres if water and spray as a foliar spray during evening hours.
2. Care should be taken that the undiluted cattle urine can cause scorching effect on plants.

h. Garlic Decoction

This spray can be prepared if there is a sudden incidence of pest and no sprays are available with the farmer to manage the pest.

Materials required

Garlic- 1 kg and Water – 10 litres.

Method

1. Boil 1 kg of garlic in 10 litres of water separately.
2. Allow the solution to cool and sieve the solution with a muslin cloth.

3. Dilute one part of the solution with 10 parts of water and spray on the crop during evening hours.

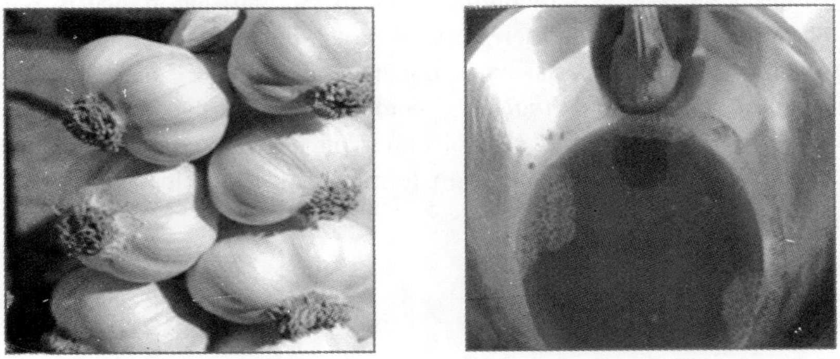

Figure 30: Garlic decoction preparation.

i. Lime Sulphur Spray

Boling lime and sulphur together makes lime sulphur. The mixture is used as a dormant spray of fruit trees to control diseases such as blight, anthracnose, powdery mildews and some insects including scales, thrips and mites.

Materials required

Quick lime –2kgs; Sulphur-4 kg and Water – 20 Litres.

Method

1. Boil the water in an iron / zinc /tin vessels. Do not use copper vessel.
2. Add lime to the boiling water and allow it to boil for 5 min.
3. Make a paste of sulphur with water and add it slowly to the boiling water containing lime.
4. Continue boiling for an hour; maintain the water level by addition additional quantities of water.
5. When the greenish yellow solution turns amber colour, stop boiling and allow the solution to settle down for a day.
6. Store the orange red coloured supernatant liquid in an airtight container.
7. One part of the solution is diluted 40 times with water and sprayed on crops.

Figure 31: Lime sulphur spray.

C. Microbial pesticides or bio-pesticides

The use of microorganisms as bio-control agents is gaining importance in recent years. Bio-pesticides are the living organisms or their derived parts which are used as bio-control agents to protect the crops against the insect pests. Entomopathogenic viruses of baculovirus group, bacterial insecticides, particularly *Bacillus thuringiensis,* entomo-fungal pathogens, protozoans and insect parasitic nematodes have been found to control important pests of crops. These bio-pesticides are commercially available and are quite difficult to formulate in the field conditions.

a. Types of Microbial Bio-pesticides

1. Bacterial Bio-pesticides.
2. Fungal Bio-pesticides.
3. Viral Bio-pesticides.

b. Method of application of bio-pesticides

a) Seed Treatment

1. Prepare 5% jaggery solution by boiling 500 gm of jaggery in 10 litres of water for 15-20 minutes. Depending on the quantity of seeds to be treated sufficient solution has to be prepared.
2. The solution has to be cooled.
3. Open the pack of bio-pesticide and mix the contents in the above solution. The general recommendation is 10gm of bio-pesticide/kg of seed.
4. Heap the seeds to be treated on a polythene sheet and the bio-pesticide solution is poured over the seeds and mixed thoroughly.
5. Dry the seeds in shade and sown immediately.

b. Nursery bed

1. One kilogram of bio-pesticide is mixed with 100 kilograms of good quality soil. Forest soil can also be used. In areas where forest soil is not available, 30 kg of well rotten dung can be mixed with 70 kg of soil to prepare a good soil mixture.
2. The soil mixture can be used as a nursery soil or it can be mixed with the nursery bed of size 10 square metres.

c. Soil drenching

1. Prepare a solution by adding 10 gram of bio-pesticide in a litre of water and stir the solution well.
2. Using a rose can to drench the soil with the solution.

d. Seedling dip

1. Prepare a solution by adding 10 grams of bio-pesticide in a litre of water.
2. Dip the seedlings in this solution for 30 minutes.

e. Soil application

1. A mixture is prepared by adding 5 kilograms of bio-pesticide and 100 kilograms of well rotten cattle dung or FYM.
2. Broadcast this mixture on soil immediately before ploughing. On a standing crop, after the application of this mixture, light irrigation has to be given.

f. Foliar spray

One kilogram of bio-pesticide is mixed in 150 litres of water and sprayed on the crop. The solution is sufficient for one acre.

D. Biological control Agents

The biocontrol agents are categorized into two groups:

a) Predators

b) Parasitoids

Predators

Predators are larger insects or animals which capture and feed on other insects or animals. They are bigger in size with well developed sense organs and require more than one prey for their survival. The insects are divided into 29 orders and out of them insects belonging to 16 orders comprise of predaceous members. Including the spiders and mites, there are more than 200,000 species of predators. Most of the crops attract a large number of predators. The use of synthetic pesticides is the major threat to these predators. When the number of predators decline, managing the pests become a major problem. It is estimated that in any given crop there are more than 300 to 500 species of predators. Spiders represent the largest and the most diverse group and used in biological control. Also mites are used as predators. The general characteristics of predators are:

a) capture and feed more than one prey for their survival.

b) relatively large size compared to the prey.

c) predaceous on larvae and adults of pests.

d) more active than the prey.

Most of the insect predators which are reared in the laboratory are released at a rate ranging from 600 per acre (*e.g.* Rodalia beetle for the control of pest cotton cushion scale) to 7 lakhs per acre (*e.g.* Chrysoperla for the control of aphids).

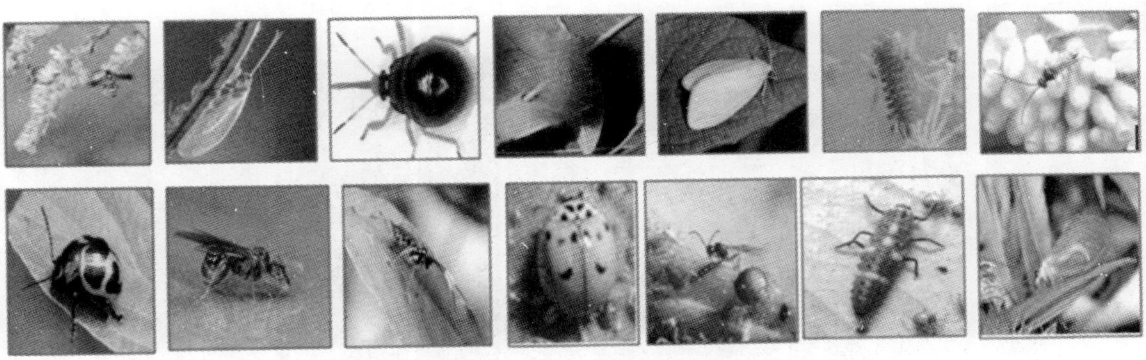

Figure 32: Different biological agents.

Table 2: Examples of predators and corresponding prey.

Predators	Prey
Lady bird beetles	Aphids scale insects and mealy bugs
Ground Beetles	Many soil dwelling insects
Rove beetles	Many soil dwelling insects
Tiger beetles	Various insects on the ground surface
Lace wings	Aphids
Hover flies	Aphids
Robber flies	Various insects
Wasps	Caterpillars of many insects
Spiders	Many flying adult insects
Predatory mites	Plant feeding mites
Toads	Various soil inhabiting insects
Lizards	Various soil inhabiting insects
Birds	Various soil inhabiting insects

The insect parasitoids are those whose larvae feed internally or externally on the body of another insect. The attacked insect is called as host and sustains the parasite larvae throughout its development. At times insect parasitoids are slightly smaller than their hosts wherein a single parasitoid larvae develops inside each individual host insect. However, when the parasitoid is very small in size, many parasitoid larvae develop within the host. The parasitoids are specific to the stage of the insect that they attack. Some attack only larva and are called as larval parasitoids, some attack on pupa which are called as pupal parasitoids and those attack on eggs are called as egg parasitoids. The egg parasitoids like *Trichogramma sp*, *Telenomus sp*, *Tetrastichus sp* Damage the egg stage of the pest by preventing the larval stage to appear which causes the damage to the crops. *Trichogramma sp* is the egg parasitoid produced in large amounts in the laboratory for the control of pests. It controls bollworms, paddy stem borer, sorghum stem borer, fruit borers etc. The general recommendation of parasitoids is about 3000 to 50,000 adults per acre.The pupal parasitoids like *Tetrastichus israeli* controls pests like American boll worms, paddy leaf rollers, many caterpillars in their pupal stages.

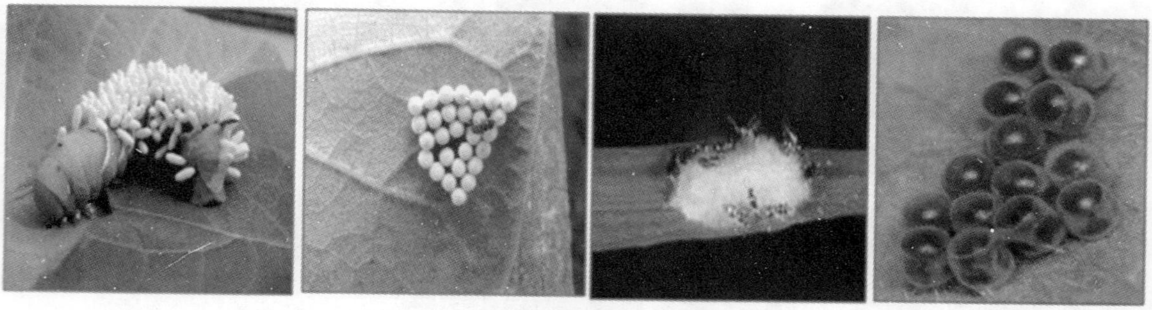

Figure 33: Different parasitoids.

Biological control is an important component in organic farming in managing pests. Their handling is safe not only to the farmers who are using them but also to the environment unlike synthetic pesticides. The most interesting fact is that the biocontrol agents are very specifi c to the pests, which refl ects its success in controlling pests. On the other hand if organic agriculture is practiced, over a period of time, these natural enemies would be in the fi eld always and can benefi t the farmer to a very large extent.

E. Botanical Pesticides

The use of plant and plant products to manage pests are referred as botanical pesticides. It is estimated that about 2121 plant species possess pest control properties; 25 of these plant species posses the characteristics required for an ideal botanical insecticide. The extracts prepared from the above listed plants have insecticidal, antifeedant, repellant antifungal, antiviral and antibacterial properties. Following are the plants most widely used for the pest management.

Table 3: Use of botanicals as biopesticide.

Plant		Protocol	Effective against
Adathoda vesica (Malabar nut tree)	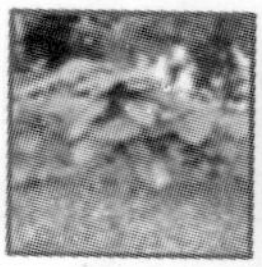 	One kg of fresh leaves is finely pounded+six litres of water and allowed to soak for a day. Liquid is then filtered and a suitable sticker, such as local soap	acaricidal / antifeedant
Curcuma longa (Turmeric)	 	1 kg of dried turmeric (fine powder) + 3 to 4 litres of cow urine. One litre of the solution is diluted in 10 litres of water and sprayed	tomato wilt /semi loopers, mites/ green leaf hoppers/ aphids
Asafoetida (Hing)	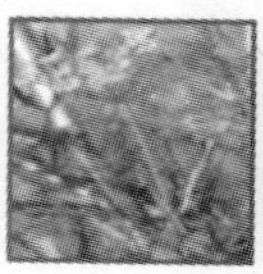 	40–50 gm of asafoetida is tied in a cotton cloth and placed in the irrigation channel /About 100 to 150 gm of asafoetida (per litre of water) are boiled for 10 to 15 minutes. After the solution is cooled, it is sprayed with 40 to 50 litres of fresh water on crops.	termites / american boll worm
Allium cepa (Onion)		2–3 kg of common salt and 3–4 kg of onion are mixed for every 100 kg of pulses chickpea (Cicer arietinum) seed in storage. can also be used as a foliar spray.	storage pests
Capsicum annuum (Chilli)	 	4–5 dry chilies per kilo of seed	storage pests

Ricinus communis (Castor)	 	500 gm of shelled castor seed or 750 gm of unshelled seed crushed +two litres of water. Heat the contents for for ten minutes. Solution is filtered and then ten litres of water is added	
Nicotiana sp (Tobacco)		One kilogram of tobacco leaves are boiled in 10 litres of water for 30 minutes. The solution thus obtained after boiling is diluted 20 times and sprayed on crops.	quick acting
Lantana camara	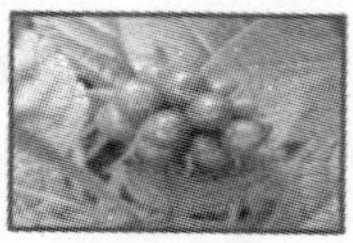	Whole plant/dried leaves and flowers	Aphids
Coriandrum sativum (Coriander)		200 gm of crushed seeds+ one litre of water. Contents are boiled for ten minutes filtered, and diluted at a rate of 1 : 2 with water	Spider mites

F. Plants for Insect Management

There are many plants which can be used along with the main crops to deter the pests. These are the simple and low cost technologies which a farmer can adopt as a tool to limit the pest attack on crops. There can be many traditional practices and knowledge of local plants which have pest deterrent qualities.

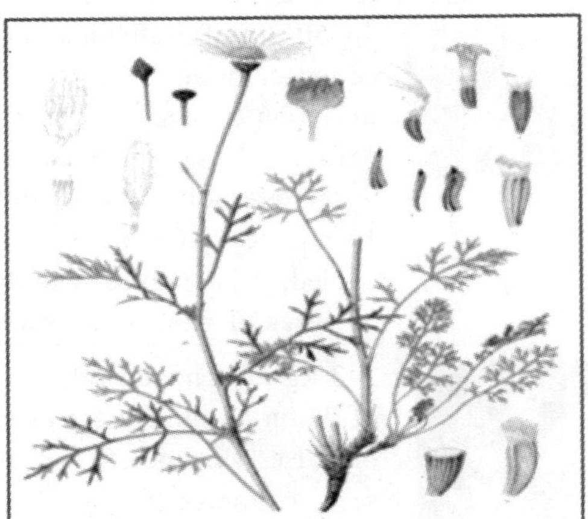

Figure 34: Pyrethrum plant.

5. Seed and Seed Treatment Techniques

Quality of the seed is very important to obtain good germination and optimum plant density. To produce quality seeds, production techniques and appropriate seed management methods are very important. In organic agriculture open pollinated seeds play a very important role than hybrid seeds. The hybrid seeds are expensive and can be used for only one season; their seeds cannot be conserved and used again. Farmers have to be fully dependent on the seed companies every season which increases the risk especially for small holders. Open pollinated varieties breed true to type and adapt well to lowinput agriculture using organic principles. The advantages of open pollinated seeds are:

- It insures seeds for the small scale farmers for their next season.
- Traditional varieties are not adapted to industrial chemical farming but to small scale farming which rely on crop rotation, crop diversification and systematic varietals mix up of crops from different genetic make up.
- They are characterized by polygenic resistance to pest and diseases being genetically diverse (multi gene resistance) and ensures the farmer against pest outbreak, crop losses, biotic and abiotic stresses.
- They contain more micro nutrients which combat malnutrition and micro nutrient deficiency in the diet of rural/poor people. They are suited to home gardening which is at the forefront of the struggle against malnutrition and can provide approximately half of the nutritional requirement for the marginalized population.

Figure 35: Seed treatment in practice.

A. Seed treatment with locally available resources

Seed treatment is very important and has many benefits like:

1. To prevent the spread of soil borne and storage pathogens which cause plant diseases.
2. For seed hardening against drought and some shocks.
3. To improve germination.
4. To prevent the attack of soil borne insects on the sown seeds

There are different methods available wherein the locally available materials can be used for treating seeds.

a) With Cow Urine

It facilitates germination and prevents seed borne diseases.

Method

1. Prepare cow urine solution by diluting one part cow urine in 5 parts of water.
2. Soak the seeds in the cow urine solution for 15 minutes.
3. Seeds are dried in shade and sown immediately.

b) With Cow's milk

It prevents yellowing of leaves and leaf spot diseases.

Method

1. Prepare a solution by mixing one part of cow's milk with 5 parts of water.
2. Soak the seeds in the above solution for 30 minutes.
3. Seeds are dried in shade and sown immediately.

c) With Wood ash

It prevents seedling rot.

Method

1. Prepare a solution by mixing 10 grams of ash (approximately 2 table spoons) in one litre of water.
2. Dip the vegetable seeds in the above solution for 15-30 minutes.
3. Seeds are dried in shade and sown immediately.

d) Hot water treatment

It controls seed borne pathogens in seeds which are stored for a long period.

Method

1. Boil the water till it reaches 55°C. Check the temperature with a thermometer.
2. The seeds are dipped for 7-10 minutes.
3. The treated seeds are shade dried and then sun dried. Such seeds are stored in insect proof containers and used when required.

B. Seed treatment with Bio-fertilizers

a) Azospirillum* and *Azotobacter

These are non-symbiotic and free living nitrogen fixing organisms whichproduce growth promoting substances like indole acetic acid, gibberellins, cytokinin, vitamins and also antifungal substances. *Azospirillum* and *Azotobacter* @ 100 grams per 5 kg seed is recommended. They improve the root system thereby increasing the uptake of nutrients. They can be used in crops like paddy, maize, buckwheat and vegetables like tomato, brinjal, cauliflower, cabbage etc.

b) *Rhizobium*

It is used in leguminous crops and they fix atmospheric nitrogen symbiotically. *Rhizobium* strains are specific to the crop. It is very important that the inoculants should be used only to the recommended crop.

c) P-Solubilizers

Phosphorus is an important plant nutrient which is not readily available as it is fixed in the soil. Some heterotrophic bacteria like, *Pseudomonas spp. And Bacillus spp.* have the ability to solubilise inorganic phosphorus from insoluble sources by secreting organic acids. These microorganisms can be used for all crops and can solubilize 20-30% of insoluble phosphate in the soil.

Method of seed treatment using bio-fertilizers

1. Prepare sugar or 5% jaggery solution (boil 50 gm of jaggery in 1 litre of water) and cool the solution.
2. The solution after cooling is sprinkled over the seeds and mixed properly.
3. 100 gm of bio-fertilizer is spread over required seeds and mixed until a uniform coating is formed over the seeds (Follow the recommended dose mentioned on the bio-fertilizer pack for different crops).
4. The treated seeds are dried in shade and then used for sowing.

Precautions while using bio-fertilizers

1.The sugar or jaggery solution should be cooled before treating the seeds.

2.The bio-fertilizer should be from an authentic source.

3.The bio-fertilizer pack should be checked for date of packing, expiry date before use.

6. Farmscaping

Farmscaping is a holistic and ecological approach to increase and manage biodiversity with the goal of increasing the presence of beneficial organisms. Many pest populations can be managed by enhancing the efficacy and local abundance of the existing community of natural enemies through modification of the environment. Some of the methods in farmscaping include, the use of insect attracting plants, hedge rows, cover crops, and water reservoirs to attract and support populations of beneficial organisms such as insects, spiders, amphibians, reptiles, bats, and birds that parasitize or predate upon insect pests. Insect attracting plants or insectary plants like rye and mustard are intercropped with the main crops that not only provide pollen and nectar but also attract and maintain beneficial insects in the crop landscape. Ideal farmscape plantings provide food and shelter for beneficial organisms, suppress weeds, and grow in close proximity to the main crop without competing for light, water and nutrients. Farmscaping aims to prevent damaging levels of pests from developing in the crop ecosystem. When significant numbers of natural enemies are present at critical stages during pest development, the biological control of insects becomes effective. However, in farmscaping, beneficial organisms are considered and managed - as "mini-livestock". The natural enemies require adequate supplies of nectar, pollen and plant-feeding insects and other arthropods as food to sustain and increase their populations. Flowering plants provide many of the food resources natural enemies need. In planning the farmscape it is important to identify specific plants, planting situations, and management practices that best support populations of beneficial organisms. This requires some information-gathering and management skill on the part of the grower in association with the assistance of the nearby research centre. The benefits of farmscaping are:

- Less incidence of pests and savings in pest management costs.
- Increased habitat and wildlife diversity on the farm.
- Farmscaping is adaptable to any field conditions.
- Erosion control/soil building.
- Added value by including cut flowers and medicinal plants.

The common approach to farmscape planning comprises of:

- Listing the main crops that are planted in the particular season.
- Knowledge of key pest(s) that are known to attack the main crop.
- Determining the commonly occurring natural enemies of the pest.
- Identifying appropriate plants that attract predators and pasitoids.
- Release of biological control agents.
- Use of bio-pesticides and other formulations for pest-management.

Figure 36: Farmscaping. (Courtesy: Dr. Oliver Moore)

The most successful farmscape systems have been developed and fine-tuned by experienced farmers to fit their individual requirements. Parasitic and predatory insects use flowers and extra-floral nectaries as food sources and mating sites. The nectar, a source of liquid sugar and vitamins for beneficial insects, and is critical for optimum performance. There are many flowers that provide pollen to the beneficial insects. Pollen is a source of protein and protein is critical for egg laying. Plants that are good pollen producers are maize, sunflower, and plants of wild carrot family or Umbelliferae such as dill, fennel, caraway, coriander, and parsnip.

7. Weed Management

Organic weed management begins with careful planning of the cropping system to minimize weed problems. There is a need to utilize the biological and ecological processes in the field and throughout the agro-ecosystem to give crops the advantage over weeds. In addition, mechanical and other control measures are usually needed to protect organic crops from the adverse effects.of weeds. The tillage and cultivation can degrade soil quality and increase the risk of erosion losses, while other organic weed management tools are more soil-friendly. For example, a diversified rotation of fast growing cash crops and cover crops can enhance soil organic matter, tilth, and fertility, provided that a sufficient quantity and diversity of residues are returned to the soil to feed the soil life. Grazing livestock after the harvest of crop to can help to remove weeds and can add fertility to the soil in the form of manure. However, intensive grazing can also compact the soil. Cover crops help to manage weeds to a large extent. The cover crops can be slashed and an in situ mulch can be formed that can enhance the soil benefits of the cover crop and suppress many annual weeds compared to ploughing them back into the soil. In the absence of cover crops under certain conditions, mulching with forest litter, straw, crop residues and other locally available materials are also quite effective.

Figure 37: *In situ* mulching with weeds.

Organic Weed Management Tools

1. Crop planning, observation and innovation
2. Fast growing main crop
3. Crop rotation
4. Cover crops
5. Mulching
6. Livestock
7. Tillage, cultivation tools and implements

Organic weed management consists of many-component strategies tailored to each region, cropping system, and farm. The other factors like choice of a crop variety, planting date and nutrient management are very important in suppressing the weeds. A combination of tactics would result in better weed management through the additive, synergistic or cumulative action of methods than any one method.

CHAPTER 7

Biodynamic Agriculture

Biodynamic Agriculture

Biodynamic Agriculture is a holistic system of farming developed by Dr. Rudolf Steiner in 1924. This system of farming is based on the lines of traditional systems being practiced in India since ages. It is based on systematic and synergistic harnessing of energies from cosmos, earth, plant and cow. It includes certain dynamic practices in addition to the basic biological practices of cultivation. The inputs utilized in this system are:

i) Biodynamic (BD) preparations numbered from 500-508

ii) Liquid manures

iii) Cow Pat Pit manure (CPP)

The biodynamic preparations are numbered from 500 to 508. Preparation 500 and 501 are made by using cow dung and quartz silica. The other seven preparations are made by fermenting parts of certain herbs in a peculiar manner. These herbs are yarrow, chamomile, stinging nettle, oak bark, dandelion, valerian and equisetum respectively. These preparations 502-507 are called the compost preparations and are utilized in making the Biodynamic compost, liquid manures and CPP. These preparations activate synthesis of different nutrients in the final biodynamic compost. These are the components of biological agriculture, capable of affording long-term sustainability to agriculture and particularly to the ecosystem. Other preparations like liquid manures, CPP and biodynamic stem paste are utilized for a number of cultural practices in farming. Another unique feature in biodynamic agriculture system is the use of biodynamic planting calendar for various farm operations. Efforts are made to enliven the soil and restore its fertility in the form of humus, increase the living system of soil by skilful application of appropriate crop rotation.

A. Biodynamic Preparations

a. Preparation 500 (Cow Horn Manure)

This preparation is one of the basic formulations in biodynamic agriculture. It is commonly known as preparation 500, the biodynamic field spray. Cow horn manure is prepared by burying a cow-horn filled with fresh cow dung of a lactating cow. Such horns are buried in soil during autumn and winter period and are lifted out in the early spring in accordance with the planting calendar. In other words, the horns are buried in September-October and lifted during March-April during descending moon. The manure when knocked out of the horn gets transformed into sweet smelling colloidal humus. About 35 gms of BD 500 is required for one-acre land. The required quantity of the preparation is placed in a vessel containing good quality water and stirred rhythmically in clockwise and anticlockwise direction alternately for one hour. First, the movement is clockwise, till a vortex is formed, then the movement is reversed till another vortex is created, again the movement is reversed, and so on for one hour. The stirred preparation 500 is immediately sprayed on the land with the help of a sprayer or a broom. The broom is dipped into the bucket containing stirred preparation 500 and is splashed on the desired area.

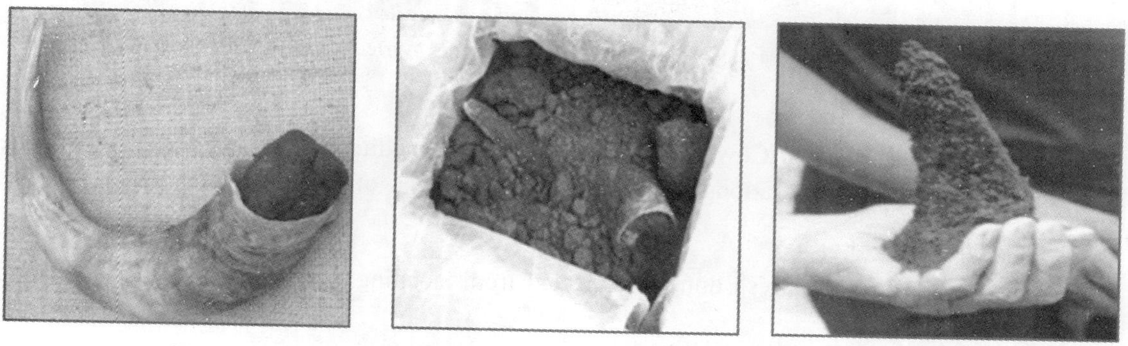

Figure 38: Cow horn manure.(Courtesy: Biodynamics Agriculture Australia Ltd)

b. Preparation 501 (Horn Silica)

This is made from a crystal of quartz (Silicon oxide, SiO_2). The crystal is ground up into a fine powder as smooth as talcum powder. It is moistened with water, put into a cow horn and buried in the soil during spring and summer. It is important to have good quality quartz crystals, well formed and clear, which should allow the light to flow right through. A good clear crystal will cause the refraction of light in the same way as the prism.

Horn Silica Preparation

Preparation 501 is also stirred in a similar manner like preparation 500. It is stirred for one hour, using one gram in 13.5 litres of water and sufficient for one acre land. It is applied in the morning in a fine mist, using a high pressure spray, allowing the mist to be suspended briefly in the air so that the sunlight can shine momentarily through the mist on the plant, the fine mist is allowed to drift over the crop. For smaller area knapsack sprayer is used with their fine nozzles turned skywards to produce misting effect. As a general rule, 501 should be sprayed at the early stage of the plant and after the fruit set. Preparation 501 is generally sprayed in the morning during spring and early summer and sometimes in the autumn if the lushness of growth demands it. As the season advances during summer, 501 is sprayed progressively early in the morning. Because of the stimulation of the light, it is advised not to use 501 in drought conditions and should only be used once in the spring on pasture as overuse can accentuate drought conditions. Due to the enhancement of photosynthesis of the plant, the starches, sugars and cellulose are improved. The quality of grain produced is greatly increased.

Figure 39: Horn Silica manure. (Courtesy: Biodynamics Agriculture Australia Ltd)

c. Cow Pat Pit Manure (CPP)

The manure obtained from Cow Pat Pit is a very useful medium or vehicle for spreading the influence of the compost preparations 502-507 over a large area of farmland, orchard or garden.

Materials required

i) *Cow dung* – 65 kg of fresh dung is collected from lactating cow which is fed with a good quality fodder.

ii) *Egg shells* – egg shells are added into the CPP as a calcium source. About 300 gm of crushed egg shells powder is required for one pit.

iii) *Basalt rock/ granite dust* – rock dust is added as a silicon source. About 300 gm of rock dust is required for one pit.

iv) *Biodynamic preparations* – 2 sets each of biodynamic preparations 502-506 and 20 ml solution of preparation 507 is required.

Method

1. Dig a pit and line the 4 sides of the pit with bake bricks. The size of the pit with baked bricks lining should be 3 feet × 2 feet × 1 feet dimension. The bottom of the pit should not be lined.
2. A site with good drainage and not prone to dry out should be chosen. A vegetable garden is ideal.
3. Sprinkle water to turn the cow dung into dough like consistency.
4. Rock dust and crushed egg shells are sprinkled over the cow dung.

5. The bricks lining is moistened with water before placing the dung mixture.
6. Mix the dung for 10-15 minutes and place it in the pit. Care should be taken that the dung is not tightly packed.
7. Insert two sets of preparations 502-506 by making holes over the dung layer. Two sets of preparation 507 is stirred for 15 minutes in 2 litres of water and sprinkled over the cow dung from the top.
8. Place a wet jute sack over the cow pat pit to maintain moisture. The pit should be in a well shaded and aerated area.
9. After one month the manure is gently aerated with a garden fork for uniform mixing.
10. Cow Pat Pit (CPP), manure matures in 35-40 days.

Storage

The Cow Pat Pit manure is stored in earthen pot placed in a dark room. Avoid the cow pat pit from drying.

Application

Cow Pat Pit manure can be applied along with preparation 500. It requires 15 minutes of stirring. The common practice is to make a spray solution of CPP with BD 500. After 45 minutes of stirring of BD 500 solution 100gm of CPP is added to the solution and both 500 and CPP are stirred for 15 minutes. The spray solution is then sprayed. CPP is also used as foliar spray (10%) plastering the cut ends of the plants after running and in seed treatment similar to biofertilizers.

Figure 40: Preparation of cow pat pit manure.

Benefits

1. Fastest way to get the effect of all biodynamic preparations to the soil.
2. An array of beneficial microbes is build up which reduces many plant diseases.
3. Provides nutrients and stimulates plant growth.

B. Use of Planting Calendar

The life patterns of all living organisms are woven into the cosmic rhythms. The present scientific world may not accept the fact regarding the influence of the cosmic rhythms and the constellations on the life forms. Human life, as well as animal and plant life, is strongly dependent on the rhythms of the earth. The plant and animal life is instantly influenced by the sidereal and synodic relationships of the sun, earth, moon and other planets. On the basis of such influences the planting calendar is prepared for agricultural operations during different timings of the year.

i) Moon opposite to Saturn

It occurs approximately once in 29.5 days.

Activities to be undertaken

1. Seed sowing, transplanting, grafting, pruning and layering.
2. Spraying BD501 (Cow horn silica) to manage pests.
3. Spraying liquid manures and foliar sprays.

ii) Full moon

This occurs every 29.5 days.

Activities to be under taken

1. Sow seeds two days before sowing
2. Apply liquid manures and CPP (Cow Pat Pit) manure
3. Spraying bio pesticides to control pest and disease.
4. Drench the animals for internal parasites (48 hours before)

iii) New moon

This happens once in 27.5days

Activities to be carried out

1. Avoid sowing seeds
2. Cutting timber

iv) Ascending periods

The moon moving in an arc from east to west and when this arc gets higher everyday, the Moon is ascending.

Figure 41: Planting calendar.(Courtesy: Biodynamic Association of India, Tamil Nadu)

Activities to be undertaken

1. Sowing of crops
2. Spray BD501
3. Spray liquid manures and CPP 68

_v_) Descending periods

The moon moving in arc from East to West and this arc gets lower everyday, the Moon is said to be in descending phase.

Activities to be carried out

1. Transplanting of seedlings.
2. Spraying BD500 (Cow horn manure).
3. Making and spreading compost.
4. Pruning tress.
5. Land preparation activities.

_vi_) Nodes

These are the days when moon pass the sun's path. It creates negative influences on the growth of plants.

Avoid all agricultural activities during nodes.

vii) Apogee

Moon's orbit around the Earth is an elliptic. The point where the moon is far away to earth is called Apogee.

Activities to be undertaken

1. Planting potatoes
2. Irrigating the field

viii) Perigee

Moon moves around the earth in an elliptical path. The point where the moon is closest to earth is called as perigee.

Activities to be undertaken

Spray bio-pesticides to manage pest and disease.

_ix_) Seed and fruit days

These days influence the growth of seed and fruit crops and good for sowing and harvesting for the same *e.g.* paddy, wheat, brinjal, ladyfinger and tomato.

x) Root days

These days influence growth and development of root crops and good for sowing and harvesting for the same. *e.g.* potato, carrot and beet root.

xi) Flower day

These days influence on growth and development of flowers and good for sowing and harvesting for the same. *e.g.* cut flowers, cauliflower, rose and jasmine.

xii) Leaf days

These days help in the growth and development of leafy vegetables and good for sowing and for harvesting. *e.g.* green leaves, cabbage.

CHAPTER **8**

Support Systems

Support Systems

Organic produce is a fast-growing segment of the food industry and is increasingly drawing the attention of farmers, governments and development agencies. Organic farming as a systematized and certifiable approach to agriculture is a relatively new phenomenon. It is no surprise that its adoption faces some challenges among both farmers and the public sector. In addition to a sound policy in organic agriculture, different support system needs to be nurtured and builtin for the smooth functioning of the program. Some of the successful support systems are discussed below.

Bio-village

A Bio Village, the latest in social and cultural innovation to meet the challenges of local agriculture in the twenty first century, is a village or territory where communities of people harmoniously coexist with nature to foster a bio-diversified ecological balance. Bio-village is an important concept which seeks to promote an efficient and sustainable use of natural and biological resources leading to the continuous and steady growth of off-farm and on-farm production while protecting and improving the environmental capital stock of the village. The aim is to strengthen and upgrade rural food, energy and livelihood security, both ecologically and economically. A great emphasis is placed upon appropriate, environmentally friendly and resource conserving eco-technologies that improve livelihoods and help encourage the upstart of local bio-friendly organizations. Bio-villages promote an integration of people's indigenous technical knowledge with advances in biological and communication technologies so collective social mobilization, microfinance, capacity building, multiple livelihood engineering, marketing skills, value addition, standardization of quality control, information management etc. are some of the essentials of the process. Bio-village is of, for and by the farmers with a main objective of brining prosperity through low-cost agriculture and innovative management of community needs.

Figure 42: A typical bio village.

Features of Bio-village

Main components of the bio-village are:

i) Food, shelter and other basic needs of the community will be surveyed by a group determined by a village committee .

ii) A strategy to meet the total needs of the village by the village committee will be prepared with involvement of Technology Facilitating Partners which is a consortium of researchers, government agencies and NGOs.

iii) Focus of the farmer will be farming-system involving animal husbandry and crops (grain, fruits, fiber, medicinal and aromatic plants etc.) with maximum use of locally available natural resources.

iv) Resources needed for crop and animal production will be produced in the village.

v) Organic agriculture approaches for production and protection of crops and animals will be adopted by the farmers who would be provided timely training.

vi) The different sustainable agriculture approaches and their success will serve as demonstration fields worth visiting by farmers from other areas.

The interested farmers of other villages after seeing the results and discussion with the farmers of the bio-village would become interested to adopt the same in their villages. The knowledge and technology transfer will have a ripple effect and without any assistance the spread would be quick and effective. Bio-village will have a positive impact in the socio-economic aspects of the farming community.

Farmer Field Schools (FFS)

Farmers are the best educators to the other farmers. The farmer to farmer extension helps to a great extent for information exchange and dissemination. FFS is a concept wherein the farmers of the village or a group of nearby villages meet once a week at a particular place which can be at one of the farmers house or a community hall. The farmers share their practical experiences or new techniques like:

- Soil and water conservation techniques.
- Production of organic inputs like compost by utilizing the village refuse.
- Planning for planting of trees.
- Kitchen gardens.
- Family health care gardens with medicinal herbs.
- Biogas or alternate energy sources.
- Apiculture.
- Seed production techniques.

Figure 42: Farmer field school.

Basics of a typical farmer field school

- The field is the primary resource for learning.
- The FFS meeting place is close to the study plots.
- FFS educational methods are experiential, participatory and learner-centered.
- Each FFS meeting includes the following activities: recapitulation of the previous week's session, the agro-ecosystem analysis, a special topic, insect zoo, a group dynamics activity and evaluation of the day's session and planning for the following week.
- All FFS include a field in which farmers collect information and materials to learn about, develop observation skills and practice making informed management decisions.
- A pre- and post-test are conducted as part of every field school.
- Preparation meetings precede a FFS to determine needs and develop a learning contract.
- Final meetings of FFS often include plans for follow-up activities.

Interesting experiences of the farmers are shared which are practical, simple and economical. Since the farmers are sharing their experiences there is trust amongst the fellow farmers who would attempt them in their own fields.

Participatory Guarantee Systems (PGS)

PGS is a quality assurance initiative that is locally relevant, emphasize the participation of stakeholders, including producers and consumers and operate outside the frame of third party certification. As per IFOAM definition "Participatory Guarantee Systems are locally focused quality assurance systems. They certify producers based on active participation of stakeholders and are built on a foundation of trust, social networks and knowledge exchange". PGS is a process in which people in similar situations (in this case small holder producers) assess, inspect and verify the production practices of each other and collectively declare the entire holding of the group as organic.

Components of PGS

- Shared Vision
- Participatory
- Transparency
- Trust: Integrity-based approach
- Learning Process
- Horizontality

Features of PGS

- Norms conceived by the stakeholders.
- Grassroots organization.
- Suitable to smallholder agriculture.
- Principles and values that enhance livelihoods.
- Documented management systems and procedures.
- Mechanisms to verify farmer compliance.
- Mechanisms to support farmers.
- Farmers' pledges.
- Seals or labels as evidence of organic status.
- Clear, pre-defined consequences for non-compliance.

PGS certification is only for farmers or communities that can organize and perform as a group within the village or in close-by villages with continuous territory and is applicable on, on-farm activities comprising of crop production, processing and livestock rearing and off-farm processing .Individual farmers or group of farmers having less than 5 members are not covered under PGS. They either have to opt for third party certification or join the existing PGS local group.

CHAPTER 9

Problems in Organic Farming

Organic agriculture has been growing rapidly in recent years, according to the reports available from different agencies. Yet organic food has garnered an extraordinary amount of attention from the media in the recent past. There's just one problem with this: agriculture has moved away from small-scale, local, and organic farming because these types of farms are land- and labor-intensive and don't do a very good job of feeding lots of people. In addition, they are *not* definitively better for the environment, and their growth would lead to higher food prices than most wealthy people are willing to pay.

Industrial agriculture has its costs, and buying from farmers' markets or local farms carries with it various aesthetic appeals. It's generally possible to find types of produce outside the mainstream agricultural system, such as that are too fragile for grocery stores, which select crops for hardiness and long shelf lives. Also, for some people, the idea of eating fruits and vegetables that are grown on a small plot of land with just seeds, sun, water, and care is attractive, and some farmers are carving out profitable niches in the market by catering to those impulses.

Problems in introducing organic farming

- Absence of scientific ecosystematic knowledge and its consistent application, Chemistry and other new technologies are often still regarded as "modern" and progressive;
- A lack of culturally adapted contents in agricultural extension services and institutions regarding organic production methods and marketing, Insufficient investment in ecological agricultural research (*e.g.* in the development of locally adapted varieties for food cultures);
- Not enough experience in considering and passing on indigenous knowledge.
- Market economy structures opposed to organic farming methods (lack of internalisation of external environmental costs that are caused by non-organic production methods);

- Additional distortion of market prices by government subsidies being provided for external inputs.

Organic food are always expensive than conventional food

Certified organic products are generally more expensive than their conventional counterparts (for which prices have been declining) for a number of reasons:

- Organic food supply is limited as compared to demand;
- Production costs for organic foods are typically higher because of greater labour inputs per unit of output and because greater diversity of enterprises means economies of scale cannot be achieved;
- Post-harvest handling of relatively small quantities of organic foods results in higher costs because of the mandatory segregation of organic and conventional produce, especially for processing and transportation;
- Marketing and the distribution chain for organic products is relatively inefficient and costs are higher because of relatively small volumes.

As demand for organic food and products is increasing, technological innovations and economies of scale should reduce costs of production, processing, distribution and marketing for organic produce.

Prices of organic foods include not only the cost of the food production itself, but also a range of other factors that are not captured in the price of conventional food, such as:

- Environmental enhancement and protection (and avoidance of future expenses to mitigate pollution). For example, higher prices of organic cash crops compensate for low financial returns of rotational periods which are necessary to build soil fertility;
- Higher standards for animal welfare;
- Avoidance of health risks to farmers due to inappropriate handling of pesticides (and avoidance of future medical expenses);
- Rural development by generating additional farm employment and assuring a fair and sufficient income to producers.

References

Francis, C and J. van Wart (2009) "History of Organic Farming and Certification" In Organic farming: the ecological system. American Society of Agronomy. pp. 3-18.

http://ncof.dacnet.nic.in/Training_manuals/Training_manuals_in_English/Cert_and_Inspection_manual.pdf

http://agritech.tnau.ac.in/org_farm/orgfarm_index.html

Rakshit,A (2014) Integrated Nutrient Management : Current Issues and Future Challenges In Sustainable Agriculture in 21st Century, Regal Publications, New Delhi,239-249.

Sarkar, N C A K Makar and Rakshit ,A (2008) "Accreditation and Certification on Organic Farming In Organic Farming: A Global Perspective by ICFAI University Press, Hyderabad. pp 63-100

Stinner, D.H (2007) The Science of Organic Farming. In William Lockeretz. *Organic Farming: An International History*. Oxfordshire, UK & Cambridge, Massachusetts: CAB International (CABI). pp. 40–72. ISBN 978-0-85199-833-6.

Thimmaiah,A (2010)Training Manual On Low Cost Organic Agriculture. National Organic Program ,Ministry of Agriculture, Royal Government of Bhutan, 69pp.

Appendix

Table 1: Fungal biofertilizers available in India.

Product	Ingredient	Remedy	Dosage	Company
		Phosphate Solubilizing Microorganisms		
Grotop PSB Powder	Phosphate Solubilizing Microorganisms *(Bacillas sp)* Powder 10^7-10^9 cfu/gm Liquid 10^9 cfu/ml	Solubilize unavailable organic & inorganic forms of phosphorus (80%)	**Seed**: @ 5-10 gm/kg seed **Soil**:@ 0.5-1 kg /acre along with 40-50 kg manure **Foliar**:@3ml/L water	MD Biocoals Pvt Ltd, Haryana
Mani Dharma's biopromoter	*Bacillus megatarium +Aspergillus niger*	Biopromoter facilitates root formation and plant growth. It improves soil quality with subsequent uses. It should not be mixed with antibacterial agents and inorganic fertilizers.	**Foliar**:@ 200 gms / 200mL of 'rice kanji' or 5% jaggary water.	ManiDharma Biotech Private Limited, Tamil Nadu
Multiplex Nalapak	Homogenous mixture of *Azotobacter + Azospirillium+* phosphate solubilizer+ potash mobilizer	It produces amino acids, vitamins and growth promoting substances like IAA, GA and Cytokines, which help in better growth and development of crop plants.Improves physical, chemical and biological properties of the soil.	**Soil**:@ 500 ml / 5 kg / acre along with 100kg kg well decomposed farm yard manure **Foliar**:@ 10g/ liter water	Multiplex Bio-Tech Pvt. Ltd., Karnataka
Ambiphos	Phosphate solubilizing microorganism (*Aspergillus niger)*	PSM secrete organic acids, which dissolve unavailable phosphate into soluble form and make it available to the plants.	**Foliar**:@3-5ml/L water	Ambika Biotech & Agro Services, Madhya Pradesh,

Biophos	*Bacillus megaterium var phosphaticum* (liquid formulations)	PSM secrete organic acids, which dissolve unavailable phosphate into soluble form and make it available to the plants.	**Seed**: @ 5-10 ml /kg of seed / seedling @ 125 – 250 ml in 25 –50 lit of water as coating, sett treatment @ 125-250 ml in 60-80 lit /ha for 30 minutes **Soil**: @ 500- 625 ml/ ha mixing with 250-375 FYM	Biotech International Limited, Delhi
BioP-P	Phosphate solubulizing microorganism (2×10^8 CFU/g)	PSM secrete organic acids, which dissolve unavailable phosphate into soluble form and make it available to the plants.	**Foliar**: @3-5ml/L water	Sundaram Overseas cooperation, Gujrat
PSM	Phosphate Solubilizing Microorganisms	PSM secrete organic acids, which dissolve unavailable phosphate into soluble form and make it available to the plants.	**Soil**: @ 0.5-1 kg /acre along with 40-50 kg manre	Shree Biocare India, Shree Biocare Solution Pvt Ltd, Gujrat
PSM	Phosphate solubilizing micro-organisms	PSM secrete organic acids, which dissolve unavailable phosphate into soluble form and make it available to the plants.	**Soil**: @ 0.5-1 kg /ha along with 100 kg manure	KRIBHCO, UP
		Enriching Compost		
Multiplex Sagar (Compost Poly Culture)	Homogenous mixture of *Azospirillium+ Trichoderma+ Pleuratous*	Nitrogen fixing potential, biopesticidal activity,enhances compost degradation	**Soil**: @ 1 kg /acre along with 1000kg kg well decomposed farm yard manure **Foliar**: @ 100 ml / 10 ltr water	Multiplex Bio-Tech Pvt. Ltd., Karnataka
Enriched compost Culture	*Trichoderma harzianum+ Aspergillus+ Penicillium*	Enhances compost degradation	**Foliar**: @ 100 ml / 10 ltr water	Organic Biotech Pvt Limited, Maharastra
Bio manure Culture	*Trichoderma harzianum+ Aspergillus*	Enhances compost degradation	**Foliar**: @ 100 ml / 10 ltr water	Uno Natural and Greens Private Limited, Tamilnadu
Ligno Biocom-postculture	*Trichpderma resei, Phanerochaete chrysosporium and Aspergillus awamori*	Enhances compost degradation	**Foliar**: @ 75-100 ml / 10 ltr water	Peak Chemical Industries Limited

Product	Ingredient	Remedy	Dosage	Company
		Phosphate mobilizing biofertilizer		
ManiDharma VAM	Vesicular Arbuscular Endomycorrhiza (VAM)	Solubilize phosphate and supplies P and Zn, Mn,Fe,Cu,Co,Mo to the plants. Increases the plant vigor by inducing drought resistance of young seedlings.Protects the plants from the fungal pathogens	**Soil**:@200 gms/m^2 or 2 - 5 gm/seedling ;50-200 gms/trees;3 – 5 kgs/acre (in 2 - 3 cm depth).	Mani Dharma Biotech Pvt Ltd, Tamilnadu
Ecorhiza-VAM/ Nurserrhiza-VAM	Arbuscular mycorrhiza	Improved uptake of nutrients, root development in plant and growth	**Soil**:@3-5 Kg / Acre with the 200-250 kgs of FYM/VC one tablet / plant in 2-4 inch deep hole near the plant root.	TERI, New Delhi
Root Care	Mycorrhiza (Glomus Intraradices)	Improved plant health, uptake of nutrients and reduction of environmental stress.	**Soil**:@5 Kg / Acre.	Ambica Biotech, MP
Mycorrhizae	10^5/kg Propagules with carrier material: Talc powder/ Vermiculite	Mobilize major nutrients like, phosphorus and potassium and certain micronutrients like zinc, calcium, etc.	**Soil**:@5-10 kgs of VAM per hectare is recommended along with the 1000 kgs of FYM/VC; 25 gms / plant.	Dr. Rajan Laboratories, Tamilnadu
JOSH Super/JOSH plus	Mycorrhizal (Glomus intraradices)	Root development in plant and hence growth	**Soil**:@5 kg per acre for all crops 60 I.P./ gram (infective propagules)	Cadila pharmaceuticals Limited, Gujrat
Shubhodaya	Vesicular Arbuscular Mycorrhiza (with three species-1-Isolated from desert and adoptable to harsh water strained condition. 2-Adoptable to water lodging conditions 3-Adaptable to general and acidic/basic soil conditions.)	Improved uptake of nutrients	**Soil**:@5 -10 kg per acre	COSME BIOTECH, Goa

Product	Ingredient	Remedy	Dosage	Company
TARI VAM	Vesicular Arbuscular Mychorrhiza	Mobilize nutrients like, phosphorus and zinc	**Soil**:@5 -10 kg per acre	TARI Biotech, Tamilnadu
BioCarry	Vesicular Arbuscular Mycorrhiza (VAM)	Helps in efficient mobilization & uptake of fertilizers and other nutrients by plants.	**Soil**:@5 kg per acre	Sundaram Overseas Operation, Gujrat
Symbion VAM	Solid formulations of Arbuscular Mycorrhiza	Improved uptake of nutrients, root development in plant and growth	**Soil**:@5 kg per acre	T. Stanes and\ Company Limited, Tamilnadu
CAMBAY's VARDHAK	Arbuscular Mycorrhiza (Powder & Tablets)	Helps in efficient mobilization of nutrients	**Soil**:@5 kg per acre; one tablet / plant	Neesa Agritech Private Limited Ahmedabad, Gujarat
Mycorrhiza–VAM	Arbuscular Mycorrhiza	Mobilize nutrients like, P, Zn, Cuand B	**Soil**:@10 kg per acre	KCP Sugar and Industries Corporation Ltd
Mycorrhiza-AM Biofertilizer	Arbuscular Mycorrhiza	Helps in efficient mobilization of nutrients and improved plant growth	**Soil**:@5 kg per acre	Majestic Agronomics Pvt. Ltd., HP
Colonizer	Arbuscular Mycorrhiza (powder)	Colonizes living plants roots. Improves phosphorus uptakeand imparts stress and disease resistance.	**Soil**:@2-3 kg per acre	Krishidhan Seeds Pvt. Ltd, Maharastra

Table 2: Bacterial biofertilizers available in India.

Product	Ingredient	Remedy	Dosage	Company
		Rhizobium		
Rhizobium	*Rhizobium* (CFU minimum 5×10^7 cell/g of powder, granules or carrier material or 1×10^8 cell/ml of liquid.)	• Rhizobium increases 20-25% of crop yeild • Fixes 40-250 Kg N/ha/ year atmospheric N_2 • Increase of yield, production of healthy seeds • Early root development	**Legumes**: red gram, pea, lentil, black gram, green gram and cowpea Lentil, Bengal-gram and Fodder legumes. All types of beans, methi, alfalfa, lupin, barseem etc. Oilseed legumes like soybean and groundnut.	Junna Life Sciences Private Limited Nagar, Kukatpally Hyderabad - 500072, Andhra Pradesh, Indi
Kisan *Rhizobium*	*Rhizobium* (CFU minimum 5×10^7 cell/g of powder, granules)	• Fixes 50-300 Kg. N/ hactre. • Leaves residual nitrogen for succeeding crop. • Increase yield by 10-35% • Maintains soil fertility • Rhizobium helps in accumulation of micronutrients like P, K, Ca, Mg and Fe	**Pulse Legumes** : Gram, Peas, Lentil, Moong, Urd, Cowpea, Arhar. **Oil Legumes** : Groundnut and Soyabeans. **Fodder Legumes** : Barseem, Lucerne. **Forest Legumes** : Subabul, Shisam, Shinsh.	National Fertilizers Ltd.,Corporate Office, Noida-201301 (U.P)
Sunrise *Rhizobium*	*Rhizobium* (CFU minimum 1×10^8 cell/ml of liquid.)	• Can fix 50-100 kg of nitrogen per hectare in one Crop season • Leave behind 20-30 kg. N/ha for the succeeding crop • Increase yields from 15-20%. • Rhizobium make endophytic association with non leguminous plants such as rice, wheat, maize, barley, millets and other cereal and increases crop yeild	**Seed treatment**:@100 ml / 500 gm of Rhizobium is mixed with 500ml. of Gum Arabica to make slurry. The seeds required for an acre are Mixed in the slurry and Shade dried for 30 minutes, before sowing. The treated seeds should be sown within 48hrs.**Soil**: Mix 250 ml / 2 kg of Rhizobium in 100 kg dried farm yard manure/ Annapurna and broadcast in One acre of land.	Multiplex Bio-Tech Pvt. Ltd., Karnataka
Rhyzobium (RHZ)	*Rhizobium* (CFU minimum 5×10^7 cell/g of powder, granules or carrier material)	• Inoculants are known for their ability to fix atmospheric nitrogen in symbiotic association with plants forming nodules in roots (stem nodules in *Sesabania rostrata*).	**Soil**: Mix 250 ml / 2 kg of Rhizobium in 100 kg dried farm yard manure/ Annapurna and broadcast in One acre of land.	KRIBHCO A-10, Sector-l, NOIDA Gautam Budh Nagar U.P

Product	Ingredient	Remedy	Dosage	Company
		Azotobacter		
Kisan Azotobactor culture	*Azotobacter chroococcum and Aztobacter brasilense* (Carrier: Lignite (moisture: 30-40%) Active ingredient: 1×10^7 cfu/gm)	• Fixes 20-40 mgN/g of C-Source • Produces growth promoting substances like vitamin B groups, Indole acetic acid and Giberellic acid. • 15-25% increase in yield. • Maintains soil fertility. • Biological control of plant diseases by suppressing some plant pathogens like aspergillus, fusarium, alternaria etc.	Wheat, Jowar, Barley, Maize, Paddy, Mustard, Sunflower, Sesamum. Cotton, Sugarcane, Banana, Grapes, Papaya, Watermelon, Onion, Potato, Tomato, Cauliflower, Chilly, Lady finger, Rapeseed, Linseed, Tobacco. Mulberry, Coconut, Spices, Fruits, Flowers. Plantation crops, Forest plants	National Fertilizers Ltd.,Corporate Office, Noida-201301 (U.P)
Azab	*Azotobacter chroococcum* (CFU minimum 5×10^7 cell/g of powder, granules or carrier material or 1×10^8 cell/ml of liquid.)	▪ The bacteria is known to produce growth-promoting substances like IAA, GA, cytokinins and vitamins, which have been beneficial effects on crop growth. ▪ Confer an additional advantage since they are known to produce antibiotic, which suppress a variety of root pathogens. Better effects are obtained in soils having adequate organic matter. ▪ Though the nitrogen contribution to the crop by azotobacter is only 20-30 kg/ha, the production of growth promoting substance compensates its nitrogen deficiency by promoting seeding vigor and prolonged photosynthetic activity.	Bio fertilizer for several non-leguminous crops including vegetables. **Seed treatment**: @100 ml / 500 gm of Rhizobium is mixed with 500ml. of Gum Arabica to make slurry. The seeds required for an acre are Mixed in the slurry and Shade dried for 30 minutes, before sowing. The treated seeds should be sown within 48hrs. **Soil**: Mix 250 ml / 2 kg of Azab in 100 kg dried farm yard manure/ Annapurna and broadcast in One acre of land.	

Product	Ingredient	Remedy	Dosage	Company
Bioazoto	*Azotobacter chroococcum* (CFU minimum 1×10^8 cell/ml of liquid.)	▪ Eco-friendly liquid biological fertiliser formulation containing bacteria, Azotobacter which remains around the seed / seedlings and use organic carbon include sugars, organic acids, amino acids released from germinated seeds/ seedlings set in soil or root exudates and fix atmospheric nitrogen most efficiently. ▪ It also secretes growth promoting substance like vitamins, nicotinic acid, indole acetic acid, gibberellin etc and helps in better germination, root & shoot growth, increase in productivity.	**Seeds** :@ 5-10 ml /kg of seed **Seedlings**: @ 125 –250 ml in 25 –50 lit of water as coating **Sett treatment**: @ 125-250 ml in 60-80 lit /ha for 30 minutes before transplanting **Soil application**: @ 500-625 ml/ ha after mixing with 250-375 kg of Farm Yard Manure .	Biotech International Limited, Delhi
		Azospirillum		
Aadhar	*Azospirillum* (CFU minimum 5×10^7 cell/g of powder, granules or carrier material or 1×10^8 cell/ml of liquid.)	▪ Azospirillum helps in fixing the atmospheric nitrogen (N_2) in Non-leguminous crop plants which increases the crop yeild ▪ Azospirillum is an associative type of bacteria, living in close proximity with the root zone ▪ Azospirillum maintains its population during favorable conditions ▪ During unfavorable conditions it forms cyst which helps to tide over unfavorable conditions ▪ It fixes atmospheric nitrogen, provides 30 to 50% of nitrogen requirement of the plant	**Seedling root dip :** Mix 100 ml / 500 grams of Aadhar in 40 liters of water and dip the seedling roots for 10 to 20 minutes before transplantation. **Seed treatment**:500 gm of Rhizobium is mixed with 500ml. of Gum Arabica to make slurry. The seeds required for an acre are Mixed in the slurry and Shade dried for 30 minutes, before sowing. The treated seeds should be sown within 48hrs. **Soil**: Mix 205 ml / 2 kg of Rhizobium in 100 kg dried farm yard manure/ Annapurna and broadcast in One acre of land.	Multiplex Bio-Tech Pvt. Ltd., Karnataka

Product	Ingredient	Remedy	Dosage	Company
		▪ Azospirillum produces plant growth hormones such as auxin and cytokinin ▪ Enhances seed germination, early emergence, improves plumpness and succulence of fruits and promote root proliferation		
Biospirillum	*Azospirillum* (CFU minimum 1×10^8 cell/ml of liquid.)	▪ Contains large amount of lipid granules, which enters the cortical cells of the root and fix up atmospheric nitrogen ▪ Produces biologically active substances like vitamins, nicotinic acid, indole acetic acid, gibberellin etc. ▪ Helps in better germination, root & shoot growth, better retention of flowers, enhancement of the number, size, colour, flavour of the produce. ▪ Reduces the dosage requirement of Nitrogen upto 25 – 30 %.	**Seeds** :@ 5-10 ml /kg of seed **Seedlings**: @ 125 –250 ml in 25 –50 lit of water as coating **Sett treatment**: @ 125-250 ml in 60-80 lit /ha for 30 minutes before transplanting **Soil application**: @ 500-625 ml/ ha after mixing with 250-375 kg of Farm Yard Manure .	Biotech International Limited, Delhi
Azospirillum	*Azospirillum* (CFU minimum 5×10^7 cell/g of powder, granules or carrier material)	▪ It fixes atmospheric nitrogen, provides 30 to 50% of nitrogen requirement of the plant ▪ Azospirillum produces plant growth hormones such as auxin and cytokinin ▪ Reduces the dosage requirement of Nitrogen upto 25 – 30 %.	**Seed treatment**: 500 gm of Rhizobium is mixed with 500ml. of Gum Arabica to make slurry. The seeds required for an acre are Mixed in the slurry and Shade dried for 30 minutes, before sowing. The treated seeds should be sown within 48hrs. **Soil**: Mix 205 ml / 2 kg of Rhizobium in 100 kg dried farm yard manure/ Annapurna and broadcast in One acre of land	KCP Sugar and Industries Corporation Ltd

Product	Ingredient	Remedy	Dosage	Company
		Acetobacter		
Symbion- N	*Gluconacetobacter* (bacteria in their dormant forms are suspended in liquid formulation @ 1×10^9 cells per ml of the product.)	Symbiotic bacteria capable of fixing atmospheric nitrogen by living within the sugar plant. They are found in all parts of plant body. It is suitable for sugarcane cultivation.	**Sett treatment**: @ 125-250 ml in 60-80 lit /ha for 30 minutes before transplanting . **Soil**: Mix 250 ml / 2 kg of Rhizobium in 100 kg dried farm yard manure/ Annapurna and broadcast in One acre of land.	T. Stanes and Company Limited, Tamilnadu
Biopotash	**Frateuria aurentia** (CFU minimum 1×10^8 cell/ml of liquid.)	Bacteria, which remains around the seed / seedlings and use organic carbon include sugars, organic acids, amino acids released from germinated seeds/ seedlings set in soil or root exudates.During their growth, they mobilize potash and make it available to crops.	**Seeds** :@ 5-10 ml /kg of seed **Seedlings**: @ 125 – 250 ml in 25 –50 lit of water as coating **Sett treatment**: @ 125-250 ml in 60-80 lit /ha for 30 minutes before transplanting **Soil application**: @ 500- 625 ml/ ha after mixing with 250-375 kg of Farm Yard Manure .	Biotech International Limited, Delhi

Table 3: Name of certifying & inspection agencies with address.

S.No.	Name of certifying & inspection agencies	Address
1.	Association for promotion of Organic Farming (APOF)	Alumni Asso, Building, Hebbal, Thottumugham, Bangalore - 560024 Ph. 080-23516060
2.	Bioinspectra C/O Indocert	Bellary Road, P.O. Aluva - 683105 Cochin, Kerala State Ph.0484-2630908
3.	Indian Society for Certification of organic products (ISCOP)	Rasi Building, Ponnaiyaraja-puram Coimbatore, Tamil Nadu - 641001, Ph. 0422-2471181
4.	SGS India Pvt Ltd	162/163, 8 250, Udyog Vihar, Phase-IV, Gurgaon-122015 Ph. 0124-2399757
5.	Indian Organic Certification Agency (INDOCERT)	Thottumugham P.O. Aluva-683105, Cochin, Kerala State, India Ph. 0484-2630909
6.	LACON	Mithradham Chunangardi
7.	Skallnspection and Certification Agency	Mahalaxmi Layout, No.191, 1st Main Road, Bangalore-560086

8.	International Resources for Fair Trade (IRFD)	Sona Udyog Unit No.7 Parsi Panchayat Road Andheri (E) Mumbai 400069 Ph. 022-28235246
9.	IMO Control Pvt. Ltd	26, 17th Main HAL, 11 2nd'A' Stage, Bangalore 560008 Ph. 080-25285883
10.	One Cert Asia	Agrasen Farm Vatika Road Off Ton Rd Jaipur, Rajasthan
11.	Ecocert International	54A, Kanchan Nagar, Nakshetrawadi, Aurangabad - 413002, Maharashtra 0240-2376336
12.	National Organic Certification Association (NOCA)	Pune

Table 4: List of biofertiliser manufacturer/supplier from Private/Public and PSU sector.

Sl No	Name of Private sector unit	Address
1	Yash Krishi, Takniki Ewam Vigyan Kendra	Allahabad, UP
2	T. Stanes and Company Limited	Coimbatore, TN
3	United Agro Industries	Salem, TN
4	Super Agro (I) Pvt. Ltd	Kolkata, WB
5	SUPA Biotech	Nainital, UK
6	Sivanthi Joe Coirs	Tutikorin, TN
7	Shamli Distillery and Chemical Works	Muzaffarnagar, UP
8	Rocky Ridge Farm	Bangalore, Karnataka
9	Ramel Industries Ltd.	Barasat, WB
10	Pramukh Agri Clinic	Surat, Gujrat
11	P.J. Margo Private Limited	Bangalore, Karnataka
12	Peak Chemical Industries Limited	Siliguri, WB
13	Nimbion Organics	Chennai, TN
14	Multiplex Bio-Tech Private Ltd.	Bangalore, Karnataka
15	Laxmi Biochem	Kolkata, WB
16	J. Organics	Theni, TN
17	Iswaryam Organics	Coimbatore, TN
18	Indian Oil Corporation Limited	Mumbai, Maharasthra
19	Hindustan Petroleum Corporation Ltd.	Mumbai, Maharasthra
20	Hindusthan Bioenergy Limited	Lucknow, UP
21	Green Star Fertilisers Limited	Chennai, TN

Sl No	Name of Private sector unit	Address
22	First Milk Pvt Limited	Chennai, TN
22	MD Biocoals Pvt Ltd	Haryana
23	Mani Dharma Biotech Private Limited	Madhanandhapuram ,TN
24	Ambika Biotech & Agro Services	Mandsaur,Madhya Pradesh
25	Biotech International Limited	Delhi
26	Sundaram Overseas cooperation	Mumbai,Maharastra
27	Shree Biocare India, Shree Biocare Solution Pvt Ltd	Ahmedabad,Gujrat
28	KRIBHCO	Gautam Budha Nagar,UP
29	Organic Biotech Pvt Limited	Mumbai,Maharastra
30	Uno Natural and Greens Private Limited	Chennai,TN
31	TERI	New Delhi
32	Dr. Rajan Laboratories	Tamilnadu
33	Cadila pharmaceuticals Limited	Gujrat
34	Cosme Biotech	Goa
35	TARI Biotech	Tamilnadu
36	Neesa Agritech Private Limited	Ahmedabad, Gujarat
37	KCP Sugar and Industries Corporation Ltd	Chennai,TN
38	Majestic Agronomics Pvt. Ltd.	Delhi
39	Krishidhan Seeds Pvt. Ltd.	Mumbai,Maharastra
40	EID Parry (India) Limited Ltd	Bio-Products Division Chennai, TN
41	Bio Control Research Laboratories – A Division of Pest Control (India) Pvt.	Bangalore,Karnataka
42	Dutch Plantin Coir India Pvt Ltd	Coimbatore, TN
43	Basant Dairy & Vermicompost Production Unit	Bangalore, Karnataka
44	AWN Agro Pvt. Ltd	Ahmedabad, Gujarat
45	Agriland Biotech Limited	Vadodara, Gujarat
46	Bharat Laboratory &Biological House for Agriculture	Dhule, Maharashtra
47	Pyrites, Phospahates and Chemicals	Amjhore, Bihar
48	Godavari Fertilizers &Chemicals Ltd.	Secundrabad, AP
49	Lakshmi Bio-techs	Cuddalore, TN
50	Ecosense Labs (I) Pvt. Ltd.	Goregaon, Maharashtra
51	Micro Bac India,	Shyamnagar, WB

Sl. No.	Name of Private sector unit	Address
52	A.V.S. Agro Products	Ahmednagar, Maharashtra
53	Nav Maharashtra Chakan oil Mills Ltd.	Pune, Maharashtra
54	Nafed Bifertilizer	Indore, MP
55	Ajay Bio-tech India	Pune, Maharashtra
56	Kumar Krashi Mitra Bio Products	Pune, Maharashtra
57	Nitrofix Laboratories	Calcutta, WB
58	The Sima Cotton Dev. &Research Association	Coimbatore, TN
59	Niku Bio-research Lab	Pune, Maharashtra
60	Kisan Agro Chem	Dhanegaon, Maharashtra
61	Magnum Associates	Chennai, TN
62	Monarch Biofertilizers and Research Centre	Chennai, TN
63	Samarth Bio Tech Ltd	Hubli, Karnataka
64	K-Ferts Lab	Nanded, Maharashtra
65	Samruddhi Agrotech	Pune, Maharashtra
66	Maharashtra Bio-tech Industries	Pune, Maharashtra
67	Vasantdada Sugar Institute	Pune, Maharashtra
68	HMG Biotech Pvt. Ltd.	Shimoga, Karnataka
69	Bio Agro Fertilizers	Pune, Maharashtra
70	Madras Fertilizers	Manali, TN
71	Terra Firma Biotechnoilogies Ltd.	Hubli, Karnataka

Sl. No.	Name of Public sector undertakings	Address
1.	KRIBHCO	Hazira, Gujarat
2.	IFFCO	MLN Farmers Training Institute (IFFCO), Phulpur, UP
3.	NFL	Bio-Fertilizers Plant at Vijaipur, AP
4.	MFL	Manali, TN
5.	FACT	TPD Plant, Cochin, Kerala
6.	RCF	Rashtrya Chemicals & fertilizers Ltd., Mumbai, Maharashtra

Sl. No.	Name of Public Sector	Address
1	Bio-fertilizer Production unit, Deptt. of Agriculture	Salem, TN
	National Biofertilizer Development Centre	Ghaziabad, UP
2	M.P.State Cooperative Oil seed Growers' Federation	Dhar, MP
3	Main Bio Control Research Laboratory (TN Cooperative Sugar Fedn.)	Changalpattu, TN
4	Micro Biological laboratory	Pattambi, Kerala
5	College of Agriculture, Marathwada Agricultural University	Parbhani, Maharashtra
6	University of Agricultural Sciences	Dharwar, Karnataka
7	Regional Soil Testing Laboratory	Rajendranagar, AP
8	Nodule Research Laboratory, BCKV	Mohanpur, WB
9	Bio Science Laboratories	Salem, TN
10	Institute of National Organic Agriculture (INORA)	Pune, Maharashtra
11	Biological Nitrogen fixation Scheme, College of Agriculture	Pune, Maharashtra
12	Rhizobium Scheme, Deptt. of Agriculture	Durgapura, Rajasthan
13	M.A.I.D.C.	Pune, Maharashtra
14	T.N. Agricultural University	Coimbatore, TN
15	Bio-fertilizer Production unit, Deptt. of Agriculture	Pudukottai, TN
16	Gujarat State Cooperative Marketing Federation Ltd.	Kankaria, Gujarat
17	Division of Microbiology	IARI, New Delhi

Table 5: Agencies involved in domestic Marketing of organic produce in India.

S.No.	Name of organization	Address
1	NAVDANYA Trust	A-60, Hauz Khas, New Delhi-110016
2	Devine Agro Industries Ltd	C-9, Anoop Nagar, Uttam Nagar, New Delhi -59
3	Devbhoomi	Rajpur Road, Dehradun, Uttaranchal
4	Back to Nature	Near Kanak Cinema, Dehradun, Uttaranchal
5	Mahrishi Ved Vigyan Vidyapeeth	Dunda (Kunshi), Uttar Kashi
6	Institute of Rura Development (IIRD	54 A, Kanchan Nagar, Nakshetrawadi, Aurangabad 413002
7	ISCON Temple	Bangalore
8	FAB India Overseas Pvt. Ltd.	B-26, Okhla Industries Area, Phase-I, New Delhi
9	ECO-Nut Health Food Shop	J's Heritage Complex, 01 Milk Union, P.T. Road, Kodai Kanal- 624101
10	Sresta ByproduCfs PvCLtd.	203, Pavani Annexes, Road No.2, Banjara Hills Hyderabad 500034, AP
11	D.R. Agro organicAS	01, Sai Nagar, Ratnagiri, Kapadqani-387620, Gujrat
12	IOCCA	951C, 15th Cross, 8th Mai Ideal Home Township Raja Radeswari Nagar, Banqalore-560 098
13	Sunstar Overseas Ltd	40 K.M. Stone, GT Kamal Road, Bahalgarh, Sonepat, Haryana
14	IITC Organic India Ltd.	A-306, Indira Nagar, Lucknow- 227 105

Organic food products exported from India include the following:

- *Organic Cereals:* Wheat, rice, maize or corn.
- *Organic Pulses:* Red gram, black gram.
- *Organic Fruits:* Banana, mango, orange, pineapple, passion fruit, cashew nut, walnut.
- *Organic Oil Seeds and Oils:* Soybean, sunflower, mustard, cotton seed, groundnut, castor.
- *Organic Vegetables:* Brijal, garlic, potato, tomato, onion.
- *Organic Herbs and Spices:* Chili, peppermint, cardamom, turmeric, black pepper, white pepper, amla, tamarind, ginger, vanilla, clove, cinnamon, nutmeg, mace.
- *Others:* Jaggery, sugar, tea, coffee, cotton, textiles.

Figure 1. Major organic brands in India.

Advantage India

India is endowed with various types of naturallyavailableorganicform ofnutrients in different parts of the country and it will help for organic cultivation of crops substantially.There is diversity in climates- 100-10,000mm rainfall, hill, desert, strong traditional farming system- crop-treeanimal, innovative farmers, vast dry lands (60% agricult1!re land), least use of chemicals. Infact, the rain fed, tribal, north east and hilly regions, of the country where negligible chemicals used are practicing subsistent agriculture for a long period.

These areas are organic by default. As regards the availability of major organic nutritional inputs (NPK) in India, the estimate of National Centre of Organic Farming, Ghaziabad is as follows:

a) Crop residue = 3.865 milliontonnes

b) Animal dung = 3.854 milliontonnes

c) Green manure = 0.223 milliontonnes

d) Biofertiliser = 0.370 milliontonnes

Besides, there is enough scope of using biodynamic preparation, vermicompost, *Amrit Pani* etc. on on-farm production basis.

Table 6: Farmers' groups/organization involved in organic buying and selling in India

Sl. No.	Organization	Name	Contact No.
1	Rapid Organic Pvt. Ltd.	Mr. Yogesh Joshi	9950489820
2	Agronic Foods Inc.	Mr. Vinod Rathi	9414020031
3	Vantage Organic	Mr. Ankit Gupta	9214050121
4	Aayam Herbal Research	Mr. Rohit Vohra	9829000096
5	Himadri Foods Pvt. Ltd.	Mr. Nitin A Vakkar	022-40561005
6	New Punjab Agro	Mr. Sanjeev Pathak	09958833334
7	Neeru Enterprises	Mr. Amrit Kapoor	0595-2350829
8	Neeru Enterprises	Mr. Chandra S. Sharma	0595-2350829
9	Bharti Herbs	Mr. Sanjay S. Sinha	9371136322
10	Cama Farms and Orchards Pvt. Ltd.	Mr. Babubhai Patel	9714729940
11	Jagriti Agro Tech	Ms. Sumita Mitra	8527277577
12	Jagriti Agro Tech	Ms. Arti Chavan	8527277577
13	Royal Eco Farms	Mr. Kamal S Rathore	9753721157
14	Veetee Fine Foods	Mr. Pramod Kumar	9996500356
15	Banyan Organics	Mr. Rahul Goswami	9354030111
16	Himachal Organic Farmers Association	Mr. Gopal Mehta	9418110660
17	Himachal Organic Farmers Association	Mr. Arun	9816531139
18	Rosebay Agro	Mr. Sunil Goyal	9414063537
19	Sanjeevani Organics	Mr. Shrikant Sharma	9412058124
20	Sanjeevani Organics	Mr. Deepak	9412058124

PERMITTED SUBSTANCES FOR THE PRODUCTION OF ORGANIC FOODS (As per CODEX ALIMENTARIUS)

Precautions

1. Any substances used in an organic system for soil fertilization and conditioning, pest and disease control, for the health of livestock and quality of the animal products, or for preparation, preservation and storage of the food product should comply with the relevant national regulations.
2. Conditions for use of certain substances contained in the following lists may be specifi ed by the certifi cation body or authority, *e.g.* volume, frequency of application, specifi c purpose, etc.
3. Where substances are required for primary production they should be used with care and with the knowledge that even permitted substances may be subject to misuse and may alter the ecosystem of the soil or farm.
4. The following lists do not attempt to be all inclusive or exclusive, or a fi nite regulatory tool but rather provide advice to governments on internationally agreed inputs. A system of review criteria as detailed in Section 5 of these Guidelines for products to be considered by national governments should be the primary determinant for acceptability or rejection of substances.

Table 7: Substances for use in soil fertilizing and Conditioning.

Substances	Description; compositional requirements; conditions of use
Farmyard and poultry manure	Need recognized by certification body or authority if not sourced from organic production systems. Factory farming 21 sources not permitted
Slurry or urine	If not from organic sources, need recognized by inspection body. Preferably after controlled fermentation and/or appropriate dilution. Factory farming sources not permitted.
Composted animal excrements, including poultry	Need recognized by the certification body or authority
Manure and composted farm yard manure	Factory farming sources not permitted.
Dried farmyard manure and dehydrated poultry manure	Need recognized by the certification body or authority. Factory farmingsources not permitted.
Guano	Need recognized by the certification body or authority.
Straw	Need recognized by the certification body or authority.

Substances	Description; compositional requirements; conditions of use
Compost and spent mushroom and Vermiculite substrate	Need recognized by the certifi cation body or authority. The initial composition of the substrate must be limited to the products on this list Sorted, composted or fermented home refuse need recognized by the certification body or authority.
Compost from plant residues processed animal products from slaughter houses & fish industries	Need recognized by the certification body or authority.
By-products of food & textile industries not treated with synthetic additives	Need recognized by the certification body or authority.
Seaweeds and seaweed products	Need recognized by the certification body or authority.
Sawdust, bark and wood waste	Need recognized by the certification body or authority, wood not chemically treated after felling from wood not chemically treated after felling wood ash and wood charcoal need recognized by the certification body or authority
Natural phosphate rock	Need recognized by the certifi cation body or authority. Cadmium should not exceed 90 mg/kg P_2O_5
Basic slag	Need recognized by the certifi cation body or authority.
Rock potash, mined potassium salts (*e.g.* kainite, sylvinite)	Should have less than 60% chlorine.
Sulphate of potash (*e.g.* patenkali)	Obtained by physical procedures but not enriched by chemical processes to increase its solubility need recognized by the certification body or authority.
Calcium carbonate of natural origin (*e.g.* chalk, marl, limestone, phosphate chalk) Magnesium rock alcareous magnesium Rock Epsom salt (magnesium- sulphate)Gypsum (calcium sulphate)	Only from natural sources/origin.
Stillage and stillage extract	Ammonium stillage excluded
Sodium chloride	Only mined salt
Aluminium calcium phosphate	Cadmium should not exceed 90mg/kg
Trace elements (*e.g.* boron, copper, iron, manganese, molybdenum, zinc) Sulphur	Need recognized by the certification body or authority.

Substances	Description; compositional requirements; conditions of use
Stone meal	
Clay (*e.g.* bentonite,perlite, zeolite)	
Naturally occurring biological organisms (*e.g.* worms)	
Vermiculite	
Peat excluding synthetic additives	Permitted for seed, potting module composts. Other use as recognized by certification body or authority. Not permitted as a soil conditioner.
Humus from earthworms and insects	
Chloride of lime	Need recognized by the certification body or authority.
Human excrements	Need recognized by the certification body or authority. The source is separated from household and industrial wastes that pose a risk of chemical contamination. It is treated suffi ciently to eliminate risks from pests, parasites, pathogenic microorganisms, and is not applied to crops intended for human consumption or to the edible parts of plants.
By-products of the sugar industry (*e.g.* Vinasse)	Need recognized by the certification body or authority.
By-products from oil palm, coconut and cocoa (including empty fruit bunch, palm oil mill effluent (pome), cocoa peat and empty cocoa pods)	Need recognized by the certification body or authority.
By-products of industries processing ingredients from organic agriculture	Need recognized by the certification body or authority.
Calcium chloride solution	Leaf treatment in case of proven calcium deficiency

Factory farming refers to industrial management systems that are heavily reliant on veterinary and feed inputs not permitted in organic agriculture.

Table 8 : Substances for plant pest and disease control.

Substance	Description, compositional requirements, Conditions for us
I. Plant and Animal	
Preparations on basis of pyrethrins extracted from Chrysanthemum cinerariaefolium, Containing possibly a synergist	Need recognized by the certification body orAuthority. Exclusion of piperonyl butoxide after 2005 as a synergist.
Preparations of Rotenone from Derris elliptica, Lonchocarpus, Thephrosia spp.	Need recognized by the certification body orAuthority.
Preparations from Quassia amara	Need recognized by the certification body orAuthority.
Preparations from Ryania speciosa	Need recognized by the certification body orAuthority.
Commercial preparations/ products of Neem (Azadirachtin) from Azadirachta indica	Need recognized by the certification body orAuthority.
Propolis	Need recognized by the certification body orAuthority.
Plant and animal oils	
Seaweed, seaweed meal, seaweed extracts, sea salts and salty water	Need recognized by the certification body orAuthority. Not chemically treated.
Gelatine	
Lecithin	Need recognized by the certification body orAuthority.
Casein	Natural acids (*e.g.* Vinegar) Need recognized by the certifi cation body or authority.
Fermented product from Aspergillus spp. Extract from mushroom (Shiitake fungus)	Need recognized by certification body or Authority
Extract from Chlorella spp Chitin nematicides Natural origin Natural plant preparations, excluding tobacco	Need recognized by certification body orAuthority.
Tobacco tea (except pure nicotine)	Need recognized by certification body orAuthority.
Sabadilla	
Beeswax	
II. Mineral	
Copper in the form of copper hydroxide, copper oxychloride, (tribasic) copper sulphate, cuprous oxide, Bordeaux mixture and Burgundy mixture	Need, prescription and application rates recognizedby certification body or authority. As a fungicide on condition that the substance be used in such a way as to minimize copper accumulation in the soil.

Substance	Description, compositional requirements, Conditions for us
Sulphur	Need recognized by certification body or authority.
Mineral powders (stone meal, silicates) Diatomaceous earth	Need recognized by certification body or authority.
Silicates, clay (bentonite) Sodium silicate Sodium bicarbonate Potassium permanganate	Need recognized by certification body or authority.
Iron phosphates As molluscicide.	
Paraffi n oil	Need recognized by certification body or \authority.
III. Micro organisms used for biological pest controls	
Micro-organisms (bacteria, viruses, fungi) *e.g.* Bacillusthuringiensis, Granulosis virus, etc.	Need recognized by certification body or \authority.
IV. Other	
Carbon dioxide and nitrogen gas	Need recognized by certification body or authority.
Potassium soap (soft soap)	
Ethyl alcohol	Need recognized by certification body or authority.
Homeopathic and Ayurvedic preparations	
Herbal and biodynamic preparations	Need recognized by certification body or authority.
Sterilized insect males	
Rodenticides	
Products for pest control in livestock buildingsand installations.	Need recognized by certification body or authority.
V. Traps	
Pheromone preparations	
Preparations on the basis of metaldehyde containing a repellent to higher animal species and as far as applied in traps.	Need recognized by certification body or authority.
Mineral oils	Need recognized by certification body or authority.
Mechanical control devices such as *e.g.*, crop protection nets, spiral barriers, glue-coated plastic traps, sticky bands.	

Table 9: Performa of Application form and Growers Contact form for Certification.

Farmer Name	Farmer Code
Farmer Address	
Buying Station	Village Name

Crop Detail

Field code number (similar to field map)	Ha (or acres)	Date of planting (Month/ Year)	Estimated yield in Kg's	Date of list use of Chemicals Month/Year		
				Fertilizer (kg)	Herbicide (Lts)	Pesticide (Lts)
Total						

Note on field situation organic crop

	Organic holding in field with multiple owners, no clear borders All owners are organic.
	Field is clearly separated from other fields by
	Other (describe)

Other fields

Crop or describe hand use	**Ha**	**Use of chemicals Yes/No**
Total ha		

I, the farmer declare that this information is correct and that I have understood the conditions for the Organic Production. I have also received a copy of the farmer's organic contract. Date: Place: Signature	Internal inspector Name Signature:

Growers Contract

Grower Name:
Grower's address: Grower Code:
Village:

1. I, the undersigned, accept to become a member of the (name of the organization) organic project, certified and controlled by (name of certification agency).
2. I promise to follow the Internal Control System (ICS) as well as the organic agricultural principles.
3. I will not use pesticides, herbicides or systematic fertilizers on any crop within my organic export field(s) or in other crops.
4. I will avoid where possible the use of any chemical substance(s) on any other crop bordering my field(s).
5. I shall endeavor to maintain the following organic principles:
   - Use of clean planting/seeding material, when available,
   - Maintain and improve soil fertility by mulching and application of organic matter, compost, manure, green manures and/or crop residues.
   - Avoid environmental degradation cutting down three unnecessarily, burning of crop remains or any other organic material; dumping of organic toxic material (such as batteries) or burning of plastic:
   - Prevent soil erosion, constructing contour borders where necessary.
6. I will commit myself to follow the organic management training programm as organized by (name of the organization)
7. If I observe violation of the organic principles, I will report this to the internal inspector or another responsible person of (organization)
8. I understand that any violation(s) of the organic principles by even a single grower will lead to the exclusion of this production or of the entire village.
9. I will allow inspection by persons authorized by (organization) and/or (certification agency).

(Organization) will provide the following support

1. Buy the organic export crop for a sustainable and transparent price including a possible organic premium (depending on market) when the export crop is of suitable quality.
2. Provide support services to the growers by way of Internal Inspectors(s) and/or consultant(s).
3. Co-ordinate the entre project including timely submission of the project for organic certification.

Location: Date:
Grower's name:
Grower's Signature:

For (organization)
Name:
Signature:

Note: Organization here can be NGO, cooperative or farmers consortia.

Table 10: Performa of Inspection report

Farm Inspection Form

Village Name Village Code

Name of Farmer Farmer's Code

Inspection Date		Date of last inspection	
Farmer Present		Time spent	
Fields Visited			
Course of visit			

CROPS

Comments, instructions, and suggestions

Rotation	Y	N	G	F	P	
Land Preparation			G	F	P	
Use of manage	Y	N				
Use of other fertilizers	Y	N				
Recognizes pests			G	F	P	
Spray when needed	Y	N				
Weed Control			G	F	P	
General score			G	F	P	Estimated yield

ANIMAL CONDITION

Comments, instruction and suggestions

Cows	G	F	P	
Goats	G	F	P	
Treatment used	G	F	P	
Manure management	G	F	P	

STORAGE

Comments, instruction and suggestions

Clean	G	F	P	
Not Contaminants	Y	N		

Conclusion (check the appropriate box.)

Farmer is performing	G	F	P	
Farmer has violated organic standards	Y	N		
Farmer must urgently improve	Y	+/–	N	
Farmer has improved since last inspection	Y	N		

Comments and questions from the farmer

I, the farmer, agree that the information given in this report is correct and complete.

Date:	Signature:

I, the field officer confirm that the information given in this report is correct and complete.

Date:	Name:	Signature:

Y-Yes, N-No, G-Good, F-Fair, P-Poor

Fig 2. Location of six regional centres and six regional centres under the National Centre of Organic Farming (Hisar, Ghaziabad, Bangalore, Nagpur, Bhubaneshwar, Imphal and Jabalpur)